Technical Report Documentation Page

1. Report No. FHWA/TX-12/0-5836-2	2. Government Accession No.	3. Recipient's Catalog No.
4. Title and Subtitle PERFORMANCE AND COST EFFECTIVENESS OF PERMEABLE FRICTION COURSE (PFC) PAVEMENTS		5. Report Date Published: February 2013
		6. Performing Organization Code
7. Author(s) Edith Arámbula, Cindy K. Estakhri, Amy Epps Martin, Manuel Trevino, André de Fortier Smit, and Jorge Prozzi		8. Performing Organization Report No. Report 0-5836-2
9. Performing Organization Name and Address Texas A&M Transportation Institute College Station, Texas 77843-3135		10. Work Unit No. (TRAIS)
		11. Contract or Grant No. Project 0-5836
12. Sponsoring Agency Name and Address Texas Department of Transportation Research and Technology Implementation Office P.O. Box 5080 Austin, Texas 78763-5080		13. Type of Report and Period Covered Technical Report: September 2008–August 2012
		14. Sponsoring Agency Code

15. Supplementary Notes
Project performed in cooperation with the Texas Department of Transportation and the Federal Highway Administration. Project Title: Performance of Permeable Friction Course (PFC) Pavements Over Time
URL: http://tti.tamu.edu/documents/0-5836-2.pdf

16. Abstract
In this project, the research team evaluated the performance of Permeable Friction Courses (PFC) over time and compared it against other types of wearing surface pavement layers. Several pavement sections including Asphalt Rubber (AR) PFCs, Performance Graded (PG) PFCs, and dense-graded Hot Mix Asphalt (HMA) were monitored over a four-year period. Non-destructive on-site measurements included noise, drainability, texture, friction, and skid. The change of these variables with time as well as the influence of traffic, binder/mixture type, aggregate classification, and climatic region was evaluated. Accident data were also gathered and analyzed on a more comprehensive number of pavement sections across Texas. All of this information was compiled in database format. In addition, when performance issues were identified, field cores were acquired for forensic evaluation. Results from the multiyear performance data analysis and previous research were used to produce guidelines and recommendations to improve the design, construction, and maintenance of PFCs.

Performance of PFCs over time was adequate. Therefore, the continued use of PFCs in Texas is encouraged. PFCs had lower overall noise levels when compared to dense-graded HMA, and AR-PFCs were quieter than PG-PFCs. With regard to drainability, the water flow values had a tendency to increase early in the life of the pavement and remain relatively constant afterward. PG-PFCs showed better drainability as compared to AR-PFCs. The amount of rainfall helped assure the continued drainability of PFCs, especially in warm climates. Texture for PFCs remained practically unchanged over time. Both AR- and PG-PFCs had superior texture and skid vs. dense-graded HMA pavements. With regard to friction and skid, sections with aggregates classified as SAC-B per the Surface Aggregate Classification (SAC) system had statistically significantly lower values as compared to those pavement employing either SAC-A or SAC-A/B aggregates. The accident data indicated that PFCs reduce the number of accidents, injuries, and fatalities on roads in Texas.

17. Key Words Porous Friction Course, Open-Graded Friction Course, Porous Asphalt, Mixture Performance, Asphalt Mixture Permeability, Noise Reduction	18. Distribution Statement No restrictions. This document is available to the public through NTIS: National Technical Information Service Alexandria, Virginia http://www.ntis.gov		
19. Security Classif.(of this report) Unclassified	20. Security Classif.(of this page) Unclassified	21. No. of Pages 254	22. Price

Form DOT F 1700.7 (8-72)   Reproduction of completed page authorized

PERFORMANCE AND COST EFFECTIVENESS OF PERMEABLE FRICTION COURSE (PFC) PAVEMENTS

by

Edith Arámbula
Associate Research Engineer
Texas A&M Transportation Institute

Cindy K. Estakhri
Program Manager
Texas A&M Transportation Institute

Amy Epps Martin
Professor
Texas A&M University

Manuel Trevino
Research Engineer Associate
Center for Transportation Research

André de Fortier Smit
Research Associate
Center for Transportation Research

and

Jorge Prozzi
Associate Professor
The University of Texas at Austin

Report 0-5836-2
Project 0-5836
Project Title: Performance of Permeable Friction Course (PFC) Pavements Over Time

Performed in cooperation with the
Texas Department of Transportation
and the
Federal Highway Administration

Published: February 2013

TEXAS A&M TRANSPORTATION INSTITUTE
College Station, Texas 77843-3135

DISCLAIMER

This research was performed in cooperation with the Texas Department of Transportation (TxDOT) and the Federal Highway Administration (FHWA). The contents of this report reflect the views of the authors, who are responsible for the facts and the accuracy of the data presented herein. The contents do not necessarily reflect the official view or policies of the FHWA or TxDOT. This report does not constitute a standard, specification, or regulation.

This report is not intended for construction, bidding, or permit purposes. The engineer in charge of the project was Amy Epps Martin, P.E. # 91053. The United States Government and the State of Texas do not endorse products or manufacturers. Trade or manufacturers' names appear herein solely because they are considered essential to the object of this report.

ACKNOWLEDGMENTS

This project was conducted in cooperation with TxDOT and FHWA. The authors thank the project director, Robert Lee, and the project advisors German Claros, Dar Hao Chen, Feng Hong, George Reeves, John Wirth, and Raphael (Ray) Umscheid for their technical guidance and assistance throughout the duration of the project. The authors thank TxDOT for the traffic control provided during field performance monitoring. In addition, Lorena Garcia, Divya Valluri, David Zeig, Rick Canatella, Shon Chang Seon, Nicholas Sweet, Denise Hoyt, Andrew Muras, and Ross Taylor from the Texas A&M Transportation Institute are acknowledged for their collaboration in obtaining and processing field data.

TABLE OF CONTENTS

Page

List of Figures viii
List of Tables xiv
Chapter 1: INTRODUCTION 1
Chapter 2: SUMMARY OF INFORMATION SEARCH 5
    Previous Research 5
    Performance 10
    Mixture Design 14
    Construction 15
    Maintenance 17
Chapter 3: EXPERIMENTAL DESIGN AND FIELD TESTING PROCEDURES 19
    On-Site Testing Setup 22
    Testing Procedures 24
Chapter 4: PAVEMENT PERFORMANCE 43
    Noise 43
    Drainability 54
    Texture 62
    Friction 69
    Skid 76
    International Friction Index 80
    Accidents 82
    Service Life 93
Chapter 5: COST-EFFECTIVENESS ANALYSIS 105
    Cost-Effectiveness Analysis Approach 105
    Description of The Research Project 110
    Attributes, Metrics, and Scores 112
    Generalized Benefit/Cost Analysis 120
Chapter 6: GUIDELINES ON DESIGN, CONSTRUCTION, AND MAINTENANCE OF PERMEABLE FRICTION COURSES 125
    Benefits of PFC 125
    Uses of PFC 125
    Cost Considerations 128
    Materials and Design 129
    Construction 132
    Maintenance 142
Chapter 7: SUMMARY 147
REFERENCES 151
APPENDIX A: PAVEMENT DATABASE 155
APPENDIX B: PAVEMENT SECTIONS USED FOR DRAINABILITY, TEXTURE, FRICTION, AND SKID MEASUREMENTS 167
APPENDIX C: PAVEMENT SECTIONS USED FOR NOISE MEASUREMENTS 179
APPENDIX D: OBSI TEST RESULTS FOR INDIVIDUAL PAVEMENT SECTIONS . 217

LIST OF FIGURES

Page

Figure 1. Improved Mix Design Method for PFCs. 14
Figure 2. Location of the Pavement Sections Included in the Experimental Design. 20
Figure 3. Texas Climatic Regions. 22
Figure 4. Schematic of the Pavement Subsections Used to Measure WFV, Texture, and Friction. 24
Figure 5. Sound Intensity Probe Showing Leading and Trailing Edges. 26
Figure 6. Single Probe Fixture with Microphone Pair in Horizontal Position 28
Figure 7. New OBSI Fixture. 29
Figure 8. Half-Inch Sound Intensity Microphone Pair. 29
Figure 9. Larson Davis 3000+ Sound Intensity Analyzer. 30
Figure 10. Chevy Malibu Test Vehicle. 32
Figure 11. Chevy Impala Test Vehicle. 32
Figure 12. SRTT Tire (top) and AWP Tire (bottom). 33
Figure 13. WFV Permeameter Dimensions (28). 34
Figure 14. WFV On-Site Test Setup. 35
Figure 15. CTM Apparatus. 37
Figure 16. Typical DFT Output. 38
Figure 17. Tire-Pavement Friction Force Diagram. 38
Figure 18. DFT Apparatus. 39
Figure 19. Two-Wheeled Trailer with Left Wheel Path Locking Tire. 40
Figure 20. Typical Graph Output Obtained after a Skid Cycle. 41
Figure 21. AWP-SRTT Correlation Using the Same Test Vehicle. 45
Figure 22. OBSI Noise Levels with Pavement Life. 46
Figure 23. Correlation between Noise Levels and Pavement Age. 47
Figure 24. Effect of Binder/Mixture Type on Noise Levels for Measurement Year 2009. 48
Figure 25. Effect of Binder/Mixture Type on Noise Levels for Measurement Year 2010. 48
Figure 26. Effect of Binder/Mixture Type on Noise Levels for Measurement Year 2012. 49
Figure 27. Effect of Climate on Noise Levels for Measurement Year 2009. 52
Figure 28. Effect of Climate on Noise Levels for Measurement Year 2011. 52
Figure 29. Effect of Climate on Noise Levels for Measurement Year 2012. 53
Figure 30. Average OBSI Noise Levels by Climatic Region. 54
Figure 31. Effect of Traffic on Drainability. 55
Figure 32. OWP Drainability Measurements throughout the Life of the Project. 56
Figure 33. OWP Drainability Measurements with Pavement Life. 57
Figure 34. OWP Drainability for Each Measurement Year with 5-Minute Impervious Field Threshold. 59
Figure 35. OWP Drainability for Each Measurement Year with 90-Second Impervious Threshold. 60
Figure 36. Effect of Aggregate Classification on Drainability OWP Measurements. 61
Figure 37. Effect of Climate on Drainability OWP Measurements 62
Figure 38. Effect of Traffic on Texture Measurements. 63
Figure 39. OWP Texture Measurements throughout the Life of the Project. 65

Figure 40. OWP Texture Measurements with Pavement Life. 66
Figure 41. Effect of Binder/Mixture Type on OWP Texture Measurements 67
Figure 42. Effect of Aggregate Classification on OWP Texture Measurements for PFC
Pavements. 68
Figure 43. Effect of Climate on Texture OWP Measurements for PFC Pavements.................. 69
Figure 44. Effect of Traffic on Friction Measurements. 70
Figure 45. OWP Friction Measurements throughout the Life of the Project............................ 72
Figure 46. OWP Friction Measurements with Pavement Life. 73
Figure 47. Effect of Binder/Mixture Type on OWP Friction Measurements. 74
Figure 48. Effect of Aggregate Classification on OWP Friction Measurements for PFC
Pavements. 75
Figure 49. Effect of AADT on Aggregate Classification and Their Effect on OWP
Friction Measurements for PFC Pavements. 75
Figure 50. Effect of Climate on Friction OWP Measurements for PFC Pavements. 76
Figure 51. SN Measurements with Time. 77
Figure 52. Effect of Binder/Mixture Type on SN. 79
Figure 53. Effect of Aggregate Classification on SN for PFC Pavements. 79
Figure 54. Effect of Climate on SN for PFC Pavements. 80
Figure 55. Relationship between SN and IFI. 81
Figure 56. Breakdown of PFC Projects by Highway Facility. 83
Figure 57. Breakdown of PFC Projects by Climatic Region. 83
Figure 58. Breakdown of PFC Projects by Population. 84
Figure 59. Accident and Rainfall Data for Texas. 85
Figure 60. PFC Accidents on Highway Facilities... 86
Figure 61. PFC Accidents by Population Zone. 86
Figure 62. Relative Breakdown of Accidents by Weather and Road Condition. 88
Figure 63. Accident Rates of CRIS Reported Wet PFC Sections... 89
Figure 64. PFC Accidents and Weather Stations in Texas. 90
Figure 65. Accident Rates of PFC Sections with Daily Rainfall Exceeding 0.1 Inch. 91
Figure 66. Accidents per Lane Mile on PFC Sections under Varying Conditions. 92
Figure 67. Pavement Section on IH 20 in Tyler. 93
Figure 68. Pavement Section on US 281 in Pharr. 95
Figure 69. Aggregate Gradation after Binder Extraction for US 281 in Pharr. 96
Figure 70. Total and Interconnected AV Distribution for Cores Obtained from US 281. 97
Figure 71. Raveling Distress on US 90 in Houston. 98
Figure 72. Aggregate Gradation after Ignition Oven for US 90 in Houston. 99
Figure 73. Total and Interconnected AV Distribution for Cores Obtained from US 90. 100
Figure 74. Raveling Distress on SL 289 in Lubbock... 101
Figure 75. Aggregate Gradation after Ignition Oven for SL 289 FR in Lubbock. 103
Figure 76. Total and Interconnected AV Distribution for Cores Obtained from SL 289. 103
Figure 77. OBSI Noise Levels Grouped by Mixture Type. 113
Figure 78. WFVs Results Grouped by Mixture Type.. 114
Figure 79. Texture Results Grouped by Mixture Type.. 115
Figure 80. SN Results Grouped by Mixture Type. 116
Figure 81. DFT Results Grouped by Mixture Type... 117
Figure 82. Correlation between B/C Ratio and Panel Rating. 123

Figure 83. PFCs in Urban Areas to Reduce Splash and Spray (San Antonio IH-35). 127
Figure 84. PFCs on Rural Interstates (Odessa District IH-10, AR Underseal followed by AR-PFC). 127
Figure 85. US 183 Brownwood District (PFC used to Remediate Bleeding Surface Treatment). 127
Figure 86. Lufkin District US 59 Cloverleaf (PFC Used to Remediate Wet Weather Accidents). 128
Figure 87. Lufkin District US 59 through Diboll. 128
Figure 88. Introduction of Fibers into Drum on IH-30 Project in Paris District. 133
Figure 89. Application of PFC on Tacked, Newly Seal Coated Surface of IH 10 in Odessa District. 134
Figure 90. Laydown Machine with Tank for Applying Tack on IH-30 in the Paris District. 135
Figure 91. Edge Clearing that Can Cause Clogging of PFC Restricting Lateral Water Flow. 136
Figure 92. MTV Paving Operation (left photo) and Areas in the Mat Where Cold Chunks Required Removal and Patching (right photo) on US 288 in the Houston District. 136
Figure 93. Windrow Pick-Up Process on US 59 in the Yoakum District with Cold Lumps Forming at the End of the Windrow. 137
Figure 94. Notched Wedge Joint Construction in PFC. 138
Figure 95. Field Water Flow Value Versus Field Compaction Effort. 140
Figure 96. A Single Roller Pass as Allowed in the Houston District (No Roller Back Up Allowed). 141
Figure 97. Cores Taken from Distressed PFC on IH-35 in San Antonio. 143
Figure B.1. Section 1 and Section 13, US 59, Yoakum. 167
Figure B.2. Section 2, SH 288, Houston. 168
Figure B.3. Section 3, US 290, Austin. 168
Figure B.4. Section 4, IH 30, Paris. 169
Figure B.5. Section 5, SH 6, Bryan. 169
Figure B.6. Sections 6 and 7, IH 20, Abilene. 170
Figure B.7. Section 8, SH 240 And Section 25 SL 473, Wichita Falls. 170
Figure B.8. Section 9, SH 6, Waco. 171
Figure B.9. Section 10, SH 71, Austin. 171
Figure B.10. Section 11 and Section 21, US 281, San Antonio. 172
Figure B.11. Section 112, SH 6, Houston. 172
Figure B.12. Sections 14 and 15, IH 37, and Section 16 US 77, Corpus Christi. 173
Figure B.13. Section 17, IH 35, Waco. 173
Figure B.14. Section 18, SH 154, Paris. 174
Figure B.15. Section 19, IH 20, Tyler. 174
Figure B.16. Section 20, IH 20, Tyler. 175
Figure B.17. Section 22, SH 6, Houston. 175
Figure B.18. Section 23, US 287, Wichita Falls. 176
Figure B.19. Section 24, US 82, Wichita Falls. 176
Figure B.20. Section 26, US 281, Pharr. 177
Figure B.21. Section 27, US 90, Houston. 177

Figure B.22. Section 28, SL 289, Lubbock.. 178
Figure C.1. Sections 1 (PFC) and 13 (REF) on US 59, Yoakum. 180
Figure C.2. PFC Pavement Section on US 59 (Section 1)... 180
Figure C.3. HMA Pavement Section on US 59 (Section 13)... 181
Figure C.4. Northbound View of US 59. 181
Figure C.5. Section 2 on SH 288, Houston.. 182
Figure C.6. View of SH 288, Section 2, Southbound... 182
Figure C.7. View of SH 288, Section 2, Subsection 2-4, Northbound. 183
Figure C.8. Detailed View of the PFC Surface on SH 288, Section 2, Northbound. 183
Figure C.9. Section 3 on US 290, Austin... 184
Figure C.10. View of US 290, Section 3, Subsection 3-1, Eastbound....................................... 185
Figure C.11. View of US 290, Section 3, Subsection 3-3, Eastbound....................................... 185
Figure C.12. Detailed View of the PFC Surface on US 290, Section 3, Subsection 3-3,
  EB. 186
Figure C.13. Section 4 on IH 30, Paris. 187
Figure C.14. View of IH-30, Section 4, Subsection 4-1, Westbound.. 187
Figure C.15. Detailed View of the PFC Surface on IH-30, Section 4, Subsection 4-1,
  WB. 188
Figure C.16. View of IH-30, Section 4, Subsection 4-5, Eastbound. 188
Figure C.17. Detailed View of the PFC Surface on IH-30, Section 4, Subsection 4-6, EB. 189
Figure C.18. Section 5 on SH 6, Bryan... 190
Figure C.19. View of SH 6, Section 5, Subsection 5-1, Northbound. 190
Figure C.20. Detailed View of the PFC Surface on SH 6, Section 5, Subsection 5-1, NB. 191
Figure C.21. Views of SH 6, Section 5, Southbound. 191
Figure C.22. Section 6 on IH 20, Abilene.. 192
Figure C.23. View of IH 20, Section 6, Eastbound. 193
Figure C.24. Detailed View of the PFC Surface on IH 20, Section 6, Subsection 6-1,
  WB. 193
Figure C.25. Section 7 on US 83, Abilene... 194
Figure C.26. View of US 83, Section 7, Subsection 7-1, Northbound. 194
Figure C.27. Detailed View of the PFC Surface on US 83, Section 7, Subsection 7-2,
  NB. 195
Figure C.28. View of US 83, Section 7, Subsection 7-4, Southbound. 195
Figure C.29. Section 9 on SH 6, Waco.. 196
Figure C.30. View of SH6, Section 9 on SH 6, Westbound.. 197
Figure C.31. View of SH6, Section 9, Subsection 9. 197
Figure C.32. Detailed View of the PFC Surface on SH 6,Waco. 198
Figure C.33. Section 11 on US 281, San Antonio. 199
Figure C.34. View of US 281, Section 11, Subsection 12T, Northbound. 199
Figure C.35. Detailed View of the PFC Surface on US 281, Section 11, Northbound. 200
Figure C.36. Section 21 on US 291, San Antonio. 201
Figure C.37. US 281, Section 21, Subsection 8T, Southbound... 201
Figure C.38. Detailed View of the PFC Surface on US 281, Section 21, Subsection 17T,
  SB... 202
Figure C.39. Section 112 on SH 6, Houston.. 203
Figure C.40. View of SH 6, Section 112, Northbound. 203

Figure C.41. View of SH 6, Section 112, Southbound. 204
Figure C.42. Detailed View of the PFC Surface on SH 6, Section 112, Southbound. 204
Figure C.43. Sections 14 (CC2) and 15 (CC1) on IH 37, Corpus Christi. 206
Figure C.44. Views of IH-37, Section 14 (CC2), Northbound. 206
Figure C.45. Detailed View of the PFC Surface on IH-37, Section 14, Corpus Christi. 207
Figure C.46. Views of IH-37, Section 15 (CC1), Northbound. 207
Figure C.47. Detailed View of the AR-PFC Surface on IH-37, Section 15, Corpus Christi. 208
Figure C.48. Section 17 on IH 35, Waco. 209
Figure C.49. View of IH 35, Section 17, Subsection NB-1. 209
Figure C.50. View of IH 35, Section 17, Northbound (Old PFC). 210
Figure C.51. View of IH 35, Section 17, Subsection SB-2 (New PFC). 210
Figure C.52. Section 19, IH 20, Tyler. 211
Figure C.53. View of IH 20, Section 19, Subsection 19-4, Westbound. 212
Figure C.54. Detailed View of the PFC Surface on IH 20, Section 19, Subsection 19-5, WB. 212
Figure C.55. Section 20 on IH 20, Tyler. 213
Figure C.56. View of IH 20, Section 20, Subsection 20-1, Eastbound. 214
Figure C.57. Detailed View of the PFC Surface on IH 20, Section 20, Subsection 20-3, WB. 214
Figure C.58. View of IH 20, Section 20, Subsection 20-3, Westbound. 215
Figure D.1. Overall Noise Level for Section 1, US 59, Yoakum. 217
Figure D.2. Frequency Spectra for Sections 1 and 13 Measured 6/7/2012. 218
Figure D.3. Overall Noise Level for Section 2, SH 288, Houston. 218
Figure D.4. Frequency Spectra for Section 2 Measured 8/23/2011. 219
Figure D.5. Overall Noise Level for Section 3, US 290, Austin. 220
Figure D.6. Frequency Spectra for Section 3 Measured 5/30/2012. 220
Figure D.7. Overall Noise Levels for Section 4, IH 30, Paris. 221
Figure D.8. Frequency Spectra for Section 4 Measured 6/18/2012. 221
Figure D.9. Overall Noise Level for Section 5, SH 6, Bryan. 222
Figure D.10. Frequency Spectra for Section 5 Measured 6/6/2012. 223
Figure D.11. Overall Noise Level for Section 6, IH 20, Abilene. 223
Figure D.12. Frequency Spectra for Section 6 Measured 6/13/2012. 224
Figure D.13. Overall Noise Level for Section 7, US 83, Abilene. 225
Figure D.14. Frequency Spectra for Section 7 Measured 6/13/2012. 225
Figure D.15. Patch on SH6, Subsection WB-1, Waco. 226
Figure D.16. Overall Noise Level for Section 9, SH 6, Waco. 226
Figure D.17. Frequency Spectra for Section 9 Measured 6/1/2012. 227
Figure D.18. Overall Noise Level for Section 11, US 281, San Antonio. 228
Figure D.19. Frequency Spectra for Section 11 Measured 6/4/2012. 228
Figure D.20. Overall Noise Level for Section 21, US 281, San Antonio. 229
Figure D.21. Frequency Spectra for Section 21 Measured 6/4/2012. 229
Figure D.22. Overall Noise Level for Section 112, SH 6, Houston. 231
Figure D.23. Frequency Spectra for Section 112 Measured 7/20/2011. 231
Figure D.24. Overall Noise Level for Section 14, IH-37, Corpus Christi. 232
Figure D.25. Frequency Spectra for Section 14 Measured 6/20/2012. 232

Figure D.26. Overall Noise Level for Section 15, IH-37, Corpus Christi. 233
Figure D.27. Frequency Spectra for Section 15 Measured 6/20/2012. 233
Figure D.28. Overall Noise Level for Section 17 (Old PFC), IH-35, Waco. 234
Figure D.29. Overall Noise Level for Section 17 (New PFC), IH-35, Waco. 235
Figure D.30. Frequency Spectra for Section 17 Measured 6/1/2012. 235
Figure D.31. Overall Noise Level for Section 19, IH 20, Tyler. 236
Figure D.32. Frequency Spectra for Section 19 Measured 6/19/2012. 237
Figure D.33. Overall Noise Level for Section 20, IH 20, Tyler. 237
Figure D.34. Frequency Spectra for Section 20 Measured 6/19/2012. 238

LIST OF TABLES

Page

Table 1. General Information of the Pavement Sections Included in the Experimental Design. 21
Table 2. T-Test Results for the OBSI Noise Levels with Respect to Binder/Mixture Type for Measurement Years 2009, 2011, and 2012. 50
Table 3. T-Test Results for the OBSI Noise Levels with Respect to Binder/Mixture Type for All Measurement Years. 51
Table 4. Classification per Climatic Region of Pavement Sections Used for Noise Measurements 51
Table 5. Statistical Differences in Noise Levels Based on Climatic Region. 54
Table 6. Pavement Sections with Statistical Significant Differences MPD OWP vs. BWP. 64
Table 7. Pavement Sections with Significant Differences in Texture OWP vs. BWP. 71
Table 8. Annual Accident Rates of Analysis Sections. 91
Table 9. Annual Injury and Fatality Rates of Analysis Sections. 92
Table 10. Forensic Test Results for US 281 in Pharr. 95
Table 11. Forensic Test Results for SL 289 FR in Lubbock. 102
Table 12. Categories and Attributes Used for the Cost-Effectiveness Framework. 107
Table 13. Friction Measurements by McDaniel and Thornton 2005 (*39*). 108
Table 14. Proposed Factors and Levels to Consider in Field Section Identification. 111
Table 15. Pavement Sections Selected for the Purpose of the CEA. 112
Table 16. Proposed Metrics for OBSI. 113
Table 17. Proposed Metrics for WFV. 114
Table 18. Proposed Metrics for Texture. 115
Table 19. Proposed Metrics for Friction Number and Skid Number. 117
Table 20. Proposed Metrics for Raveling and Bleeding. 118
Table 21. Proposed Metrics for Performance Duration. 118
Table 22. Proposed Metrics for Number of Accidents. 119
Table 23. Proposed Metrics for the Number of Wet Days. 119
Table 24. Proposed Metrics for Additional Costs and Benefits. 120
Table 25. Final List of Pavement Sections Used for the B/C Analysis. 121
Table 26. Scores for the Generalized Benefit/Cost Analysis Model Variables. 124
Table 27. Mixture Designs Used on US 59 Yoakum and US 290 Austin. 140
Table 28. Weather Event: Frost or Black Ice (*11,43*). 146
Table A.1. Database Tables. 155
Table A.2. AGG_GRAD Data Field Descriptions. 156
Table A.3. BNDR Data Field Descriptions. 156
Table A.4. CRASH Data Field Descriptions. 157
Table A.5. Weather Condition (Wthr_Cond_ID). 158
Table A.6. Light Condition (Light_Cond_ID). 158
Table A.7. Road Alignment (Road_Algn_ID). 158
Table A.8. Rural-Urban Indicator (Rural_Urban_Type_ID). 159
Table A.9. Surface Condition (Surf_Cond_ID). 159

Table A.10. Crash Severity (Crash_Sev_ID). 159
Table A.11. Road Crash Location (Road_Part_Adj_ID). 159
Table A.12. CTM Data Field Descriptions. 160
Table A.13. CTM_DROP Data Field Descriptions. 160
Table A.14. CTM_MPD Data Field Descriptions. 161
Table A.15. CTM_RMS Data Field Descriptions. 162
Table A.16. DFT Data Field Descriptions. 162
Table A.17. OBSI Data Field Descriptions. 163
Table A.18. PERFORMANCE Data Field Descriptions. 163
Table A.19. PFJ Data Field Descriptions. 164
Table A.20. RUT Data Field Descriptions. 164
Table A.21. SKID_TRAFFIC Data Field Descriptions. 165
Table A.22. WFV Data Field Descriptions. 165
Table C.1. Location Coordinates for Section 1 181
Table C.2. Location Coordinates for Section 2 184
Table C.3. Location Coordinates for Section 3 186
Table C.4. Location Coordinates for Section 4 189
Table C.5. Location Coordinates for Section 5 191
Table C.6. Location Coordinates for Section 6 192
Table C.7. Location Coordinates for Section 7 196
Table C.8. Location Coordinates for Section 9 198
Table C.9. Location Coordinates for Sections 11 and 21 202
Table C.10. Location Coordinates for Section 112 205
Table C.11. Location Coordinates for Sections 14 and 15 208
Table C.12. Location Coordinates for Section 17 211
Table C.13. Location Coordinates for Section 19 213
Table C.14. Location Coordinates for Section 20 215

CHAPTER 1:
INTRODUCTION

Permeable friction course pavements (PFCs) are being used more frequently by the Texas Department of Transportation (TxDOT) as a pavement surface layer for reasons of safety, amenity, and environmental benefits. PFCs are defined in TxDOT Specifications Item 342 as a surface course of a compacted permeable mixture of aggregate, asphalt binder, and additives mixed hot in a mixing plant (*1*). These pavements are a new generation of open-graded friction course (OGFC) designed to achieve good drainability and reduce the generation and propagation of traffic noise. PFCs also offer better ride quality based on the International Roughness Index (IRI), better vehicle handling due to the coarser surface texture, and the flexibility of application as a wearing surface over Portland cement concrete (PCC) or hot mix asphalt (HMA).

The usual percent air voids (AV) in PFCs is at least 18 percent. With such a high fraction of AV, PFCs are a superior pavement in wet weather. This and other desirable characteristics of PFCs that translate into benefits to the traveling public include (*2- 6*):

- Reduced wet-weather splash and spray.
- Reduced risk of hydroplaning and wet skidding.
- Reduced number of wet weather accidents.
- Reduced noise levels.
- Increased pavement marking visibility during heavy rain events.
- Increased resistance to rutting.
- Cleaner water runoff when compared to dense-graded HMA.

Statistics from Cedar Park Police Department in Travis County demonstrate that the number of wet weather accidents on FM 1431 went down from 23 in 2003 to two in 2004 after the road was overlaid with PFC. This reduction came despite almost twice as many rain days in 2004 (*7*).

There are also some potential disadvantages associated with the use of PFCs. The most obvious difference between PFCs vs. a conventional dense-graded HMA is the decreased expected service life (typically 6–10 years for PFCs vs. 12–15 years for HMA), although the overall construction cost remains approximately the same for the PFCs with higher binder contents but higher AV contents. The most common distress encountered in PFCs is raveling, which can diminish the *durability* of the pavement (*8*). Rapid progression often characterizes raveling in PFCs, which can be associated with:

- Placement of the mixture at low temperature.
- Inadequate compaction.
- Insufficient asphalt content.
- Asphalt binder drain-down.
- Aging of the binder.
- Binder softening generated by oil and fuel drippings from traffic accidents.

In addition to durability, the *functionality* of PFCs defined as the ability of the pavement to maintain its beneficial characteristics with regard to increased drainability and reduced tire/pavement noise should be considered. These benefits may be impacted when dust and debris accumulate and clog the AV structure of the pavement or when the PFC densifies under the effect of traffic loads. State agencies usually schedule preventive maintenance focused on cleaning the PFC pore structure and prolonging the condition of the pavement surface. In the absence of cleaning activities, the initial permeability and noise reduction capacity are expected to decrease such that, at the end of the functional life (when the functional characteristics are lost), PFC behaves more like a conventional dense-graded HMA. Other challenges associated with PFCs are related to winter maintenance. It is difficult to follow the traditional snow and ice control measures because the use of sand and small aggregates to improve the surface friction are likely to clog the AV structure, and the deicing salts usually applied on the surface of the pavement do not remain in place for long.

When considering the use of PFC as a new pavement or an overlay, individual TxDOT districts must determine the critical factors. As with HMA, there is no unique design that is suited for all locations and necessities. There is a need to monitor PFCs in order to find the best mix design that can keep the drainage and noise reduction abilities over time under different climates and traffic conditions.

Several TxDOT projects have been completed in recent years that addressed mix design and construction and maintenance of PFCs in Texas (0-5262), noise abatement from PFCs (0-5185), and winter maintenance issues (0-4834). Reports from these projects, while thorough, do not address the long-term performance capabilities of PFCs (*9,10,11*). In order to complete that evaluation, performance of PFCs and other types of pavement surfaces was tracked over time in this project to determine changes in properties. This project was initiated in 2008 with the purpose of creating a multiyear database to track several PFC performance parameters in terms of functionality, durability, and safety. *Functionality* included drainability and noise reduction effectiveness. *Durability* monitored raveling, rutting, and cracking. *Safety* was measured in terms of texture, friction, skid resistance, and accident data. This database, the contents of which are described in Appendix A, was used to provide guidelines and recommendations to improve current TxDOT PFC specifications.

This report presents in Chapter 2 a summary of information search that was conducted at the beginning of the project. Then, in Chapter 3, the experimental design and field testing procedures are described. The next chapter describes the field performance of the pavement sections in terms of functionality, durability, and safety. Chapter 5 details the cost-effectiveness analysis performed using the project data. Next, guidelines for the design, construction, and maintenance of PFCs are offered in Chapter 6. The report concludes with a summary of findings and recommendations.

CHAPTER 2:
SUMMARY OF INFORMATION SEARCH

As part of this project, a report was prepared to review current research related to mixture performance, design, construction, and maintenance of surface courses using PFCs, and outline recommendations based on the literature search. This chapter presents a summary of previous research projects, Texas A&M Transportation Institute (TTI) technical report number 0-5836-1 titled *Synthesis of Current Research on PFC Performance, Design, Construction, and Maintenance* (*12*), and other recent literature.

PREVIOUS RESEARCH

To date, several TxDOT-sponsored research projects on PFCs have been completed. A brief description of these efforts and main findings is provided next.

0-4834 Cold-Weather Performance of New Generation OGFC

This project investigated winter performance and maintenance issues of Texas PFC pavements. The project included:
- Separate national and statewide surveys of practice (including sections for use, performance, cost, maintenance, and other).
- Laboratory results for permeability and abrasion.
- Methodology for remote detection of icing.
- Recommendations for construction and maintenance based on geographic location (*11*).

The results from TxDOT Project 0-4834 not only documented the history of Texas and other states with PFC use but also established an initial reference point that could serve as a benchmark for performance of older PFC.

0-5185 Noise Level Adjustments for Highway Pavements

The goal of this project was to provide evidence to the Federal Highway Administration (FHWA) to allow the use of quiet pavements for noise avoidance and abatement on federally funded projects. The two main objectives were: a) to quantify how much of the noise generated by traffic derived from the tire/pavement interaction (and thus potentially reduced by quieter pavement) and b) to estimate how long and under what conditions a quiet pavement remained quiet. A major part of the project and data collection effort focused on building a database of noise measurements at the tire/vehicle interface and at the roadside, plus absorption spectra from test section cores, which were correlated to mix design and clogging. An attempt was made to assess the effect of aging in terms of time and traffic, but most of Texas PFC pavements at the time were less than five years old. The generated database captured the acoustic properties of many PFC pavements statewide at one or two occasions.

On-board sound intensity (OBSI) measurements showed that the average level for the typical pavements was 101 dBA, whereas the average level for the PFCs was 98.1 dBA. This represented a reduction of almost 3 dBA, which is significant in terms of equivalent traffic levels, as it corresponds to half the traffic volume, assuming that the tire/pavement interaction generated most of the roadside noise.

0-5220 Investigation of Stormwater Quality Improvements Utilizing Permeable Pavement and/or the Porous Friction Course

This study focused on the hydraulic and water quality aspects of PFC. Runoff from PFC and conventional pavement was collected at three different monitoring sites in Austin over a four-year period. The stormwater samples were analyzed for solids, nutrients, and metals. The data showed that the quality of runoff from PFC was substantially better than from conventional pavement, with a reduction in total suspended solids of 90 percent. Significant reductions were also observed for total copper, lead, and zinc, though concentrations of dissolved constituents were not significantly different.

Clogging of the PFC layers was also a phenomenon of interest, but measurements of porosity and hydraulic conductivity were fairly constant over the study period, suggesting that any clogging was occurring slowly. Drainage modeling efforts showed good agreement with field measurements and provided some of the first predictions of water depth on the surface of PFC.

0-5262 Optimizing the Design of Permeable Friction Courses

This project evaluated three main aspects of PFC mix design: volumetrics, functionality (permeability or drainability and noise), and durability. Construction and maintenance guidelines were also produced. With regard to volumetrics, different methodologies for determining G_{mb} (the vacuum method and dimensional analysis) and two methodologies for determining G_{mm}

values (measured and calculated) for use in calculation of total AV content resulted in significant differences in terms of total AV contents and optimum asphalt content. The dimensional analysis and a calculation procedure based on laboratory measured G_{mm} values at low asphalt contents (3.5 and 4.5 percent) were recommended for determining G_{mb} and G_{mm} values, respectively. In addition, two methods (the vacuum method and a methodology proposed for dimensional analysis) were evaluated to compute water-accessible AV content defined as the proportion of AV in the compacted mixture that is accessible to water. This water accessible AV content was considered as an alternative parameter for mix design and evaluation. The dimensional analysis was also recommended to compute the water-accessible AV content of PFC mixtures.

The evaluation of drainability in the field as described in Tex-246-F was recommended to guarantee adequate initial mixture drainability. The main recommendations regarding durability and aging included:

- Using the draindown test (Tex-235-F).
- Utilizing the Cantabro test in both dry and wet conditions to evaluate mixture durability and susceptibility to moisture damage with maximum losses of 20 and 35 percent, respectively.
- Specifying density requirements for field compaction to help control durability.
- Conducting further long-term field observation of PFC aging to establish a quantitative relationship between aging and the Cantabro loss test, and validate a significant impact of binder oxidation on pavement performance shown based on Cantabro loss test results.
- Directing future research efforts toward tracking PFC performance to validate the durability parameters and developing a more fundamental analytical model that provides a better understanding of PFC mixture performance and a more reliable mix design method.

Construction and maintenance guidelines were also developed as part of the project. The construction guidelines address mixture production, storage and transportation, surface preparation, mixture placement, compaction and joint construction, and mixture acceptance. These guidelines were mostly based on interviews with engineers and inspectors in TxDOT districts with experience in the construction of PFCs. The research team obtained data from published literature and from onsite field observations during the construction of 10 PFC projects around the state. One of the recommendations in the construction guidelines was to use water flow (Tex-246-F) to set the roller pattern and to ensure adequate permeability.

0-6615 Use of Fine Graded Asphalt Mixes

This ongoing research explores the use of fine open-graded friction courses (OGFC) as a new alternative to chip seals for maintaining low-volume roads. Chip seals are effectively used to keep roads sealed, but these generate excessive noise levels when used near populated areas. New Mexico has significantly improved the performance of their low volume pavements in the past 10 years by using a specific type of OGFC as an alternate to a chip seal or microsurfacing. The OGFC is finer and placed significantly thinner than TxDOT's PFC mixes (i.e., between 13 mm [0.5 inch] and 19 mm [0.75 inch]). The OGFC mixture results in a much quieter and smoother material vs. a typical chip seal or microsurfacing.

In this study, researchers performed a cost and performance comparison between this type of OGFC and a chip seal and/or microsurfacing. It is anticipated that based on the results of the study, the OGFC could be used not only as an option for surfacing low volume roads but also as a thin overlay on aged pavements that do not require strengthening. These types of overlays last longer than chip seals and are more appealing in urban environments.

0-6635 Water Quality Performance of Permeable Friction Course on Curbed Sections

The objective of this ongoing project is to extend the applicability of PFC by demonstrating that a comparable improvement in quality is also possible on highway sections with curb and gutter, allowing widespread use for highway-widening projects in urban areas where limited right-of-way makes the rural cross-section infeasible. TxDOT Project 0-5220 demonstrated that PFC provides a substantial improvement in the quality of highway runoff. This reduction in pollutant concentrations was sufficient to meet the requirements of the Texas Commission on Environmental Quality (TCEQ) for use as a permanent best management practice on the Edwards Aquifer recharge and contributing zones. However, that approval was only extended to the configuration tested, which was a rural highway cross-section with a vegetated shoulder. This project intends to develop the information necessary to receive approval from TCEQ to use PFC on highway sections with curb and gutter by installing water quality monitoring equipment to document the quality of runoff from selected test sites in the Austin area. The project is also investigating the hydraulic performance of PFC in this new configuration. In addition, the analysis is considering the effect of PFC pavement on drainage system sizing.

0-6679 Performance Life of Various HMA Mixes in Texas

The objective of this ongoing project is to define the service life of various HMA mixtures, including PFC, so that planners, designers, and maintenance personnel can use these numbers to run the life cycle cost analyses of the pavement structures and the ultimate selection of the HMA type. Knowing performance lives is also key information when developing pavement maintenance programs. Currently, the estimated performance life of different HMA mixes (including the frequency of overlay) by the designers in Texas is highly subjective. This problem will be addressed in terms of not only the properties of the asphalt mix type but also in the context of the actual pavement design process (e.g., staged construction vs. perpetual pavement), quality of construction, maintenance needs, traffic volume, and environmental conditions.

National Projects

Other projects at the national level include NCHRP Project 9-41 Performance and Maintenance of Permeable Friction Courses and PFC Testing at the National Center for Asphalt Technology (NCAT) Test Track. NCHRP 9-41 includes a worldwide literature review and a survey of local and worldwide highway agencies to gather information on design, construction, maintenance, safety, performance, and volume of use of PFC mixtures. The main objective of this project was focused on recommending guidelines for design, construction, and maintenance of PFC mixtures based on the state of the practice and literature review. With regard to PFC testing at NCAT, seven different PFC sections were constructed and tested at the NCAT Test Track, including sections W3, W4, and W5 constructed in June 2000; S4 and W8 constructed in August 2003; and N13 and S3 constructed in August and October 2006, respectively. Sections W3, N13, and S3 received approximately 10 million ESALs at the termination of testing and the remaining sections will have received approximately 20 million ESALs. Relevant data collected on these sections included sound pressure levels with the NCAT noise trailer, sound intensity levels, macrotexture measurements (sand patch and Circular Texture Meter [CTM]), skid (ASTM E 274 skid numbers and Dynamic Friction Tester [DFT] results), and roughness profiles (IRI).

PERFORMANCE

Performance of PFCs can be classified into three categories: 1) functionality, 2) durability, and 3) safety. As mentioned before, functionality includes noise reduction effectiveness and drainability, while safety can be described in terms of texture, friction, skid resistance, and number of wet weather accidents. A brief description of these variables is provided next.

Noise Reduction

It has been widely observed that use of PFCs results in noise reduction at the tire/pavement-interface. When tires are rolling over a pavement surface, air in the tire tread is compressed between the tire and the pavement, and is then released as the tires continue to roll, creating noise. However, when tires are rolling over PFCs, air in the tire treads is able to escape through the AV or pores of the mixture. Therefore, the tire/pavement generated noise can be reduced. Since tire/pavement noise on the highway dominates the overall traffic noise output, PFCs have a significant effect on reduction of overall traffic noise.

Many researchers have undertaken studies to determine the influence of different factors on the noise reduction effectiveness using different mixture types. The following conclusions were generally offered:

- Compared to all types of HMA pavement surfaces, PFCs have the lowest tire/pavement noise level.
- Tire/pavement noise levels increase with increasing vehicle speed, and the noise gradient of PFCs is around 0.17 dBA per mph.
- Lower texture and roughness of the pavement surface lead to quieter pavements.
- The sound level is more influenced by air temperature at frequencies below 1500 Hz and at temperatures below 21°C (70°F).
- The type of tires influences sound levels on the same type of pavement due to the tire treads, with studded tires producing more noise at the tire/pavement interface.
- The noise reduction effectiveness of PFCs may decrease as the pavement ages.
- The sound level increases with traffic volume due to an acceleration of pavement consolidation.
- PFCs with small maximum aggregate size generate lower noise levels.

- The noise level can be reduced by increasing the addition of crumb rubber.
- An increase in the total AV content of the mixture results in a decrease in noise levels.

Numerous studies have been conducted to determine comparative noise levels on different pavement surface types, such as PFC/OGFCs, stone matrix asphalt (SMA), and dense-graded HMA. In general, the noise reduction on PFC pavement surfaces in the United States is from 3 to 9 dBA and up to 10 dBA in Europe (*13*). Noise reduction levels reported in several international and national studies can be found in Technical Report 0-5836-1 (*12*). This information led the research team to conclude that the levels of pavement noise reduction are dissimilar due to different factors. Thereby, additional research isolating mixture properties and other pavement factors from environmental conditions needs to be pursued.

Various factors either increase or decrease the noise levels on pavements. The best noise reduction with PFCs is obtained on low volume and slow speed pavements (the minimum speed recommended for porous asphalt in Europe is 48 kph [30 mph]) located in warmer climates (where studded tires are not allowed). The types of asphalt binder commonly used in PFCs are performance grade (PG) asphalt and crumb rubber modified asphalt also known as asphalt rubber (AR). Current TxDOT specifications in Item 342 require a binder content of 5.5–7.0 percent for PG-PFCs and 8.0–10.0 percent for AR-PFCs. The aggregate gradation, nominal aggregate size, AV content, and the use of AR need to be optimized during the PFC mix design to achieve proper noise reduction.

Drainability

The drainability of PFCs is measured during pavement construction to verify compaction and throughout the life of the pavement to evaluate the evolution of water flow and estimate the functional life of the mixture. The common approach to measure the drainage capacity of porous mixtures in the field is the determination of the time of discharge of a specific water volume. Tex 246-F *Permeability or Water Flow of Hot Mix Asphalt* or other similar test procedure can be used to measure water flow in the field. In the laboratory, permeability has been measured using falling head or constant head permeameters.

Several factors can have a positive effect on drainability such as higher total AV content, reduction in compaction energy, coarser aggregate gradation, and larger maximum aggregate size. A negative impact on drainability usually occurs with higher asphalt binder content in the mixture. The total AV content measured in the laboratory is often used as an indicator of permeability/drainability; however, water-accessible AV content can be used as an indirect measure of permeability/drainability with better results.

In summary, low binder content, coarse aggregate gradation, large maximum aggregate size, and low compaction effort can yield a large AV content, which is considered directly proportional to drainability. However, decreasing the energy of compaction (in both the field and laboratory) to obtain a higher total AV content is not recommended, since durability problems can arise in mixtures with incomplete compaction due to insufficient stone-on-stone (SOS) contact. In addition, high permeability/drainability values pursued by reducing the asphalt binder content can also cause mixture durability problems. That is why TxDOT and other state agencies specify minimum asphalt binder contents for both PG-PFCs and AR-PFCs. Although permeability/drainability is somewhat accounted for in most PFC mix design procedures by specifying a minimum AV content for PFCs, research studies have suggested that this approach has limitations and additional measures need to be incorporated to ensure adequate drainability, especially during field construction (*14*).

Durability

Raveling is the most common type of distress reported as the cause of failure in PFCs (*8*). The reported service life of PFCs usually ranges from six to 10 years (*8,9,11,12*). The most influential factors affecting the durability of PFCs are the type of asphalt binder, asphalt binder content, and total AV in the mixture. The majority of agencies reporting successful application of PFCs employ modified binders in the mixtures. Recent research studies also suggest that the durability of PFCs is strongly correlated to aggregate properties (*15,16*).

In the United States, the draindown and Cantabro loss tests in dry condition are used to evaluate mixture durability. The tensile strength ratio (TSR) is used to evaluate moisture susceptibility. In order to evaluate mixture durability and susceptibility to moisture damage, the Cantabro loss test is the most appropriate test currently available for PFC mix design and laboratory performance evaluations. However, the Cantabro test may not provide enough sensitivity to become a definitive tool for selecting the optimum asphalt content of PFCs.

Safety

It has been widely observed that PFCs result in skid resistance improvement at the tire/pavement interface as compared to conventional dense-graded HMA. Because of the high correlation between skid resistance and accident rates, road engineers consider pavement skid resistance an important property for designing HMA. For example, Luce et al. (*17*) showed that when the research team used a high skid resistance mixture, they obtained a 54 percent reduction in wet weather accidents and a 29 percent reduction in overall accidents.

Numerous studies have compared the skid resistance of different pavement types—such as PFC/OGFCs, SMA, and dense-graded HMA—and concluded that PFCs have the highest pavement skid resistance. Reports indicate that an increase in average friction from 0.4 to 0.55 results in a 63 percent decrease in wet weather accidents (*18*). In addition, a study performed in Ontario, Canada, indicated that the wet weather accidents decreased by 71 percent in intersections and by 54 percent on highways by the improving the skid resistance (*19*). Several studies have looked at the influence of various factors on the skid resistance achieved by different type of pavements. The following general conclusions can be made:

- Pavement skid resistance increases with decreasing vehicle speed.
- Lower pavement texture leads to lower pavement skid resistance.
- Skid resistance increases as the air temperature decreases.
- Pavement skid resistance can have an initial increase after construction due to the asphalt binder wearing off the pavement surface. Afterward, the skid resistance tends to decrease with increasing traffic volume.
- Skid resistance increases when AR binder is used.
- Mixtures that employ aggregates with high texture levels also report higher skid resistance.

The best skid resistance is obtained in PFCs with low traffic volume, slow speed limit, and constructed in areas with moderate to cold climates. Aggregate texture, mixture texture, and the use of AR binder should be considered during materials selection and mix design to ensure adequate skid resistance. Technical Report 0-5836-1 provides a list of performance aspects to achieve a balance of desired properties for the best PFCs mixtures (*12*).

MIXTURE DESIGN

The current TxDOT PFC mix design method specifies a minimum total AV content of 18 percent to guarantee functionality, but there is no durability test included in this approach to assess the resistance to raveling of compacted PFC specimens. A recent research study developed an improved mix design method for PFCs based on the guidelines of the current method that TxDOT (20) applied. Proposed modifications, which are indicated in Figure 1 by dashed-lined boxes, included improvements in the computation of volumetric properties (density, total AV content, and water-accessible AV content) and the assessment of drainability, durability, and SOS contact. Recommendations extracted from an analysis of the effects of densification in PFCs as well as their internal structure were also integrated.

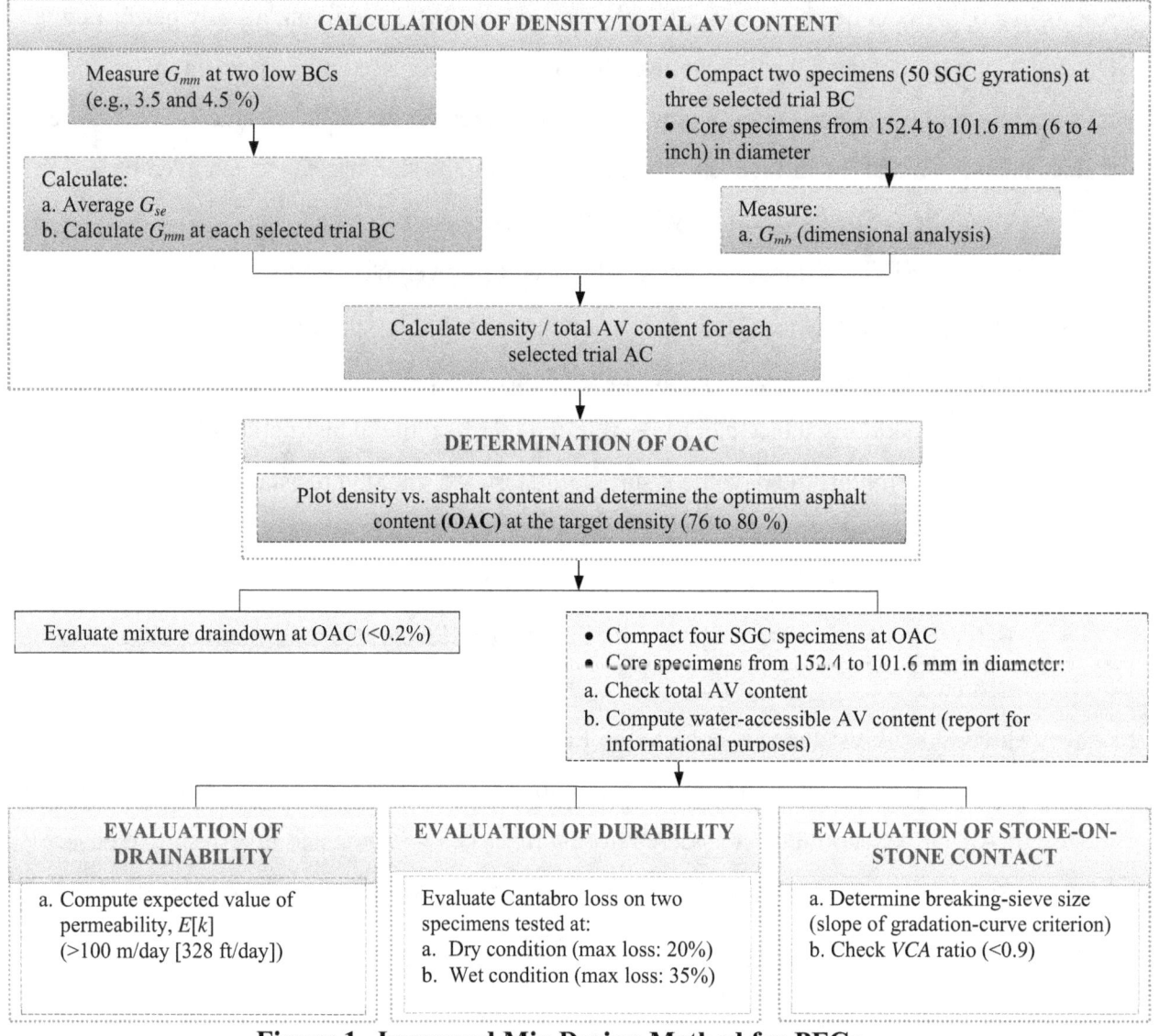

Figure 1. Improved Mix Design Method for PFCs.

CONSTRUCTION

Construction of PFCs utilizes the current techniques applied to construct conventional dense-graded HMA. However, construction of porous layers requires special considerations throughout the process as summarized below.

Mixture Production

Most PFC mixture production requires special attention to aggregate moisture control, incorporation of fibers, and use of modified binders, which is achieved by adapting conventional batch and drum plants. In addition, both the dry and the wet mixing time should be lengthened to augment fiber (mineral or cellulose) distribution when using a batch plant to produce PFC mixtures. Since PFCs are characterized by draindown susceptibility, control of mixing temperature also requires particular attention.

Mixture Storage and Transportation

Since PFCs are prone to draindown, limits on mixture storage and transportation time should be required. Tarps are necessary to avoid crusting of the PFC during transportation. Although Item 342 in TxDOT specifications does not require insulated truck beds for PFC transportation, some state DOTs do have this requirement. Preparing truck beds by using a full application of an asphalt release agent is recommended for transporting rich PFC mixtures particularly if polymer or AR binder is used.

Surface Profile

PFCs are surface layers unable to correct profile distresses or structural distress; therefore, the underlying surface should exhibit adequate conditions before placement of the PFC. Besides, lateral and longitudinal drainage of the underlying layer must be provided to ensure adequate water discharge from the PFCs. In addition, placement of PFCs over an impermeable layer is recommended to prevent problems underneath. In Texas, road engineers prefer the surface directly beneath the PFCs to be a seal coat (also known as a chip seal) and use an adequate tack coat to bond the PFCs to the underlying surface. This helps seal the surface from the intrusion of water from the surface.

Mixture Placement

The smoothness of PFCs is highly dependent on construction practices; surface depressions are more difficult to correct with PFCs than with conventional dense-graded HMA. The use of a safety edge may benefit PFCs because these pavements are daylighted at the pavement edge. Furthermore, the use of modified binders and the construction of PFCs in thin layers demands special attention to placement and compaction temperatures. Acceptable paving conditions in the United States are commonly defined as a minimum air temperature of 15°C (60°F), but it varies slightly in different states.

Material Compaction and Joint Construction

Compaction of PFCs is typically performed using static steel-wheel rollers; 8- to 9-ton tandem rollers are appropriate to complete the compaction process in thin layers. Vibratory rollers and pneumatic-tire rollers are not used during PFC compaction because they break down the aggregates and reduce the mixture drainage capacity by closing the surface pores, respectively. Keeping a maximum distance of 15 m (50 ft) between the roller and the paver is strongly recommended. Some TTI researchers have used the field water flow test (Tex-246-F) to set the roller pattern and to verify that the compacted mixture has adequate drainability.

Longitudinal and transverse joints in PFCs require special treatment since they are more difficult to construct than those found in conventional dense-graded HMA. Longitudinal joints should always be located outside the wheel paths, and longitudinal cold joints should always be avoided, if possible.

Mixture Acceptance

The practice in most agencies for mixture approval is based on the evaluation of binder content, gradation, laboratory density, and visual inspection of material variability and segregation. Adequate compaction is necessary to prevent raveling. Although a specified density in the field is not currently required, recent research recommended inclusion of a density specification for PFC construction. The density corresponding to the optimum asphalt content may be used as reference to define the density that should be required in the field. In general, all agencies specify a minimum smoothness for mixture acceptance.

MAINTENANCE

Maintenance for PFCs involves different activities than what is performed on conventional dense-graded HMA. Therefore, maintenance planning is a fundamental aspect to consider in any project involving PFCs. Maintenance of porous layers requires some special practices as summarized below.

Winter Maintenance

The high AV content and connectivity and associated thermal properties of PFCs cause earlier and more frequent frost and ice formation in this type of pavement. In fact, formation of black ice and extended frozen periods are currently considered the main problems associated with PFC maintenance in the United States. Consequently, PFCs require specific winter maintenance practices such as more frequent applications of salt (or deicing agents) than on dense-graded HMA and greater control in the homogeneous supply of deicing chemicals.

The spreading of sand to enhance friction and hasten deicing is not recommended because it contributes to the clogging of AV in PFCs. Anti-icing procedures produce the best results against black ice, freezing rain, and light snow events; deicing procedures should be reserved for events in which ice and snow have already bonded.

Surface Maintenance

Although some states using PFCs apply fog seals to perform preventive surface maintenance, there are no published reports in the United States on the application of surface maintenance for PFCs. Cleaning of PFCs such as high-pressure washing and air suction is not a common practice in the United States. However, in some European countries and Japan, different techniques are applied to clean the AV structure and maintain porosity during the pavement's lifetime. Some of these countries are also testing a two-layer drainage asphalt pavement and cleaning in order to maximize mixture functionality.

Corrective Maintenance

The use of dense-graded HMA to repair delaminated areas and potholes was indicated by all states in the United States that reported the utilization of PFC/OGFCs in 2000 (*21*). To diminish the impact of the patch on the existing PFC and facilitate water flow around it, a 45-degree rotation of the patch is recommended. Crack filling as corrective maintenance of PFCs should be avoided since it may generate drainage problems because water flow inside the mixture is diminished.

Rehabilitation

If the PFC loses its functionality by clogging, but no raveling or other major distress is present, it is still considered a functional pavement behaving like a dense-graded HMA. When major distresses are present, general recommendations and actual practices for rehabilitation of PFCs in the United States include milling and replacing of existing PFCs with new ones or other types of asphalt mixtures. However, the ideal set of technical actions for major rehabilitation of PFCs should consist of milling, recycling, and inlaying (*22*). Direct placement of new dense-graded HMA over porous mixture is not recommended because the service life of the new layer can be diminished by water accumulation inside the PFCs.

CHAPTER 3:
EXPERIMENTAL DESIGN AND FIELD TESTING PROCEDURES

In order to populate the multiyear performance database (Appendix A) and compare the performance of PFCs to other type of wearing pavement surfaces; PFCs, thin-bonded PFCs (TBPFC), ultra thin-bonded HMA (UTBHMA), and conventional dense-graded HMA pavement sections were selected. These pavement sections are located as far north as Wichita Falls to Pharr in the south and from Houston in the east to Abilene in the west (see Figure 2); therefore, all five climatic zones in the state are represented (Figure 3). There are 20 sections that remained fixed in the experimental design throughout the project as well as several rotating sections that were included year-to-year as part of forensic studies. At least four of these rotating sections were included in each performance monitoring session, with some sections being carried over for multiple years and others only being visited once. Of the 20 permanent sections, four are conventional dense-graded HMA pavement sections that serve as reference to compare changes in the performance of the PFC sections, one is a TBPFC, one is an UTBHMWC, and the rest are PFCs. The dense-graded HMA sections, which are highlighted in Table 1 and labeled with REF under mixture type, are located in different climatic zones. Details of all sections are listed in Table 1 and detailed maps of individual pavement sections including coordinates are included in Appendices B and C.

Several on-site field tests were performed to monitor drainability, texture, friction, noise level, skid resistance, and pavement distress in terms of visible signs of raveling, rutting, or cracking. Additionally, accident data for each section were obtained from the TxDOT Crash Records Information System (CRIS). TTI personnel performed field tests that included water flow value (WFV), pavement texture using the CTM, and dynamic friction using the DFT. The Center for Transportation Research (CTR) carried out noise measurements using an OBSI system. TxDOT personnel measured skid resistance and extracted from the Pavement Management Information System (PMIS) database. All data were stored in a database for performance evaluation as discussed in this report. In addition, the results of the performance monitoring, and especially the forensic studies, were used to establish guidelines and recommendations to improve current TxDOT specifications in Item 342.

Figure 2. Location of the Pavement Sections Included in the Experimental Design.

Table 1. General Information of the Pavement Sections Included in the Experimental Design.

ID	CSJ	District	County	Route	Dir	Mixture Type	Binder	Aggregate	SAC Type	Bond Type	Const M/Y	Climate
1	0089-06-076	Yoakum	Wharton	US 59	SB	PFC	PG 76-22S	Limestone	B	Seal	June 2007	WW
2	0598-02-043	Houston	Brazoria	SH 288	SB	PFC	AR	Granite/Limestone	A/B	Seal	Oct 2006	WW
3	0114-06-031	Austin	Bastrop	US 290	EB	PFC	AR	Sandstone	A	Seal	April 2007	M
4	0010-02-079	Paris	Hopkins	IH 30	WB	TBPFC	PG 76-22	Sandstone	A	Membrane	May 2006	WC
5	0049-06-061	Bryan	Robertson	SH 6	NB	PFC	PG 76-22	Sandstone/Limestone	A/B	-	May 2009	WW
6	0006-05-Xxx	Abilene	Taylor	IH 20	WB	PFC	PG 76-22 TR	Limestone	B	-	June 2005	DC
7	0033-06-097	Abilene	Taylor	US 83	NB	PFC	PG 76-22S	Limestone	B	Seal	Sept 2005	DC
8	0156-03-044	Wichita Falls	Wichita	SH 240	NB	UTBHMWC	PG 76-22	Siliceous/Limestone	A/B	Membrane	May 2008	DC
9	0049-01-085	Waco	McLennan	SH 6	WB	PFC	PG 76-22S	Limestone	B	Seal	August 2005	M
10	0265-01-103	Austin	Travis	SH 71	WB	REF	PG 76-22	Limestone/Field Sand	B	-	April 2008	M
11	0521-04-223	San Antonio	Bexar	US 281	SB	PFC	AR	Trap rock	A	Seal	May 2005	DW
112	0050-04-025	Houston	Waller	SH 6	NB	PFC	PG 76-22 TR	-	A	Seal	July 2005	WW
13	0089-08-086	Yoakum	Wharton	US 59	NB	REF	PG 70-22S	-	B	-	Sept 2004	WW
14	0074-05-089	Corpus Christi	San Patricio	IH 37	NB	PFC	PG 76-22	Limestone/Gravel	B/A	Seal	May 2004	DW
15	0074-06-197	Corpus Christi	Nueces	IH 37	NB	PFC	AR	Limestone	B	-	April 2004	DW
16	0372-01-092	Corpus Christi	San Patricio	US 77	NB	REF	PG 64-22	-	B	-	July 2009	DW
17	0015-01-164	Waco	McLennan	IH 35	NB	PFC	PG 76-22 TR	Rhyolite	A	-	May 2003	M
18	0401-01-019	Paris	Hopkins	SH 154	NB	REF	PG 64-22	Sandstone	A	-	May 2003	WC
19	0495-02-057	Tyler	Van Zandt	IH 20	EB	PFC	PG 76-22 TR	-	A	Seal	June 2008	WC
20	0910-00-085	Tyler	Smith	IH 20	EB	PFC	PG 76-22 TR	Sandstone	A	Tack	August 2009	WC
21	0073-08-150	San Antonio	Bexar	US 281	SB	PFC	AR	Sandstone/Limestone	A/B	Seal	Sept 2006	DW
22	1685-06-027	Houston	Fort Bend	SH 6	SB	TBPFC	PG 76-22S	Quartzite	A	Membrane	April 2005	WW
23	0224-01-054	Wichita Falls	Clay	US 287	NB	UTBHMWC	PG 76-22	Granite/Dolomite	A/B	Membrane	August 2005	DC
24	0044-02-072	Wichita Falls	Clay	US 82	NB	UTBHMWC	PG 76-22	Granite/Dolomite	A/B	Membrane	August 2005	DC
25	0249-11-009	Wichita Falls	Wichita	SL 473	SB	UTBHMWC	PG 76-22	Granite/Dolomite	A/B	Membrane	May 2008	DC
26	0255-08-091	Pharr	Hidalgo	US 281	SB	PFC	AR	Gravel	A	-	May 2004	DW
27	0271-09-017	Houston	Waller	US 90	WB	PFC	AR	Sandstone	A	Seal	March 2004	WW
28	0783-01-093	Lubbock	Lubbock	SL 289	WB	PFC	AR	Gravel/Limestone	A/B	Seal	Oct 2010	DC

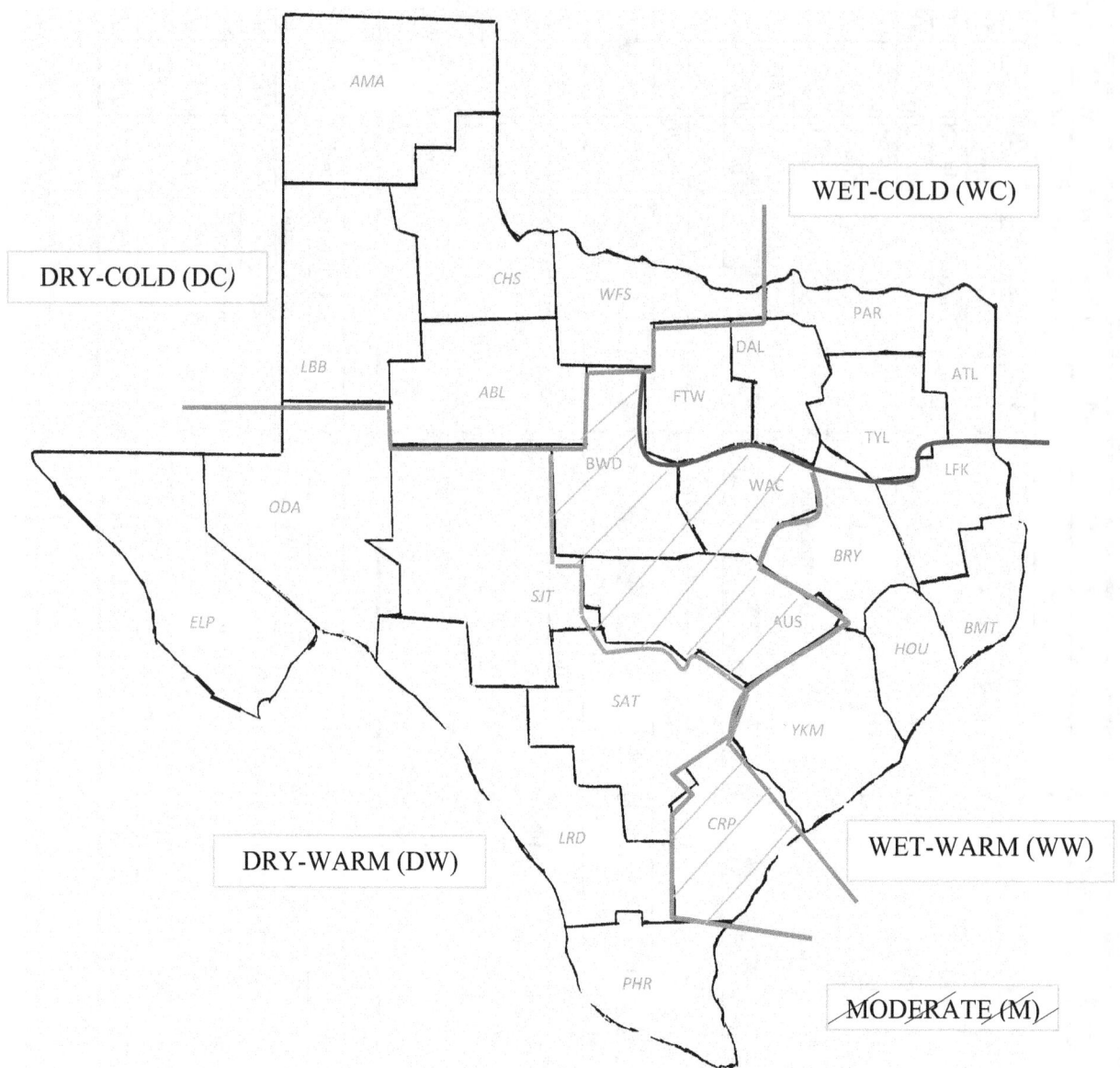

Figure 3. Texas Climatic Regions.

ON-SITE TESTING SETUP

In order to identify any differences in pavement characteristics due to the effect of traffic, the tests were performed both on the wheel path (OWP) and between the wheel paths (BWP). Most traffic lanes were approximately 3.7 m (12 ft) wide from side stripe to side stripe, not including road shoulders and/or center medians that divide traffic lanes. Each wheel path was around 0.9 m (3 ft) wide with the majority of tires passing along the center of the wheel path, thus making the wheel paths the principle area of load-induced stress. In general, the researched identified the wheel paths by looking down in the direction of travel and observing the slight rut and/or visibly different pavement texture. When testing OWP, the research team positioned the CTM and DFT where the traffic wear was the greatest, rather than favoring one side of the wheel path. The only exception was when severe cracking was present OWP; in those instances, the cracked areas were avoided to prevent skewing the results. All testing OWP was done on the right wheel path when facing the direction of travel. This was done to improve the safety of the field personnel and to put motorists at ease.

Testing BWP was done by placing the equipment near the estimated centerline of the wheel paths. In general, this area is subjected to the least amount of load-induced stress within the travel lane. BWP testing was also performed in the outer most lane of traffic. Typically, this is the rightmost lane of travel that does not include turn lanes, merge lanes, exit lanes, ramps, etc. Testing locations were always chosen to avoid testing near spots that would endanger personnel and cause undue inconvenience to motorists.

TTI personnel conducted on-site tests (i.e., WFV, texture, and friction) on three equally spaced subsections approximately 152 m (500 ft) apart (see Figure 4). Testing OWP and BWP was done in all three subsections and the average of the measurements reported as the OWP or BWP result for that pavement section. One WFV measurement was performed OWP and BWP in each subsection unless there was a reason to believe the result was inaccurate due to localized bleeding or clogging. In the case of the CTM, two replicate tests about 2 inches apart from each other are performed in each subsection. The average of the six replicate measurements was reported as the OWP or BWP texture for that pavement section. Friction was measured once in each subsection by placing the DFT in the same spot where the first texture measurement was acquired. It was important to measure texture and friction on the same location because these variables are used to calculate the International Friction Index (IFI) as will be explained later.

Figure 4. Schematic of the Pavement Subsections Used to Measure WFV, Texture, and Friction.

TESTING PROCEDURES

Noise Measurements

The most basic (and still the most accurate) method of measuring traffic noise is to set up sound pressure meters at the roadside. All other methods used to measure road noise are simply more convenient, faster ways to approximate the wayside noise via correlation. Therefore, it is essential to establish a relationship between the on-vehicle methods and the direct measurement and to check that correlation periodically during the vehicle testing. It has been observed under this project and many others, that correlation between on-board and roadside measurements are unique to the vehicle configuration (primarily tires) and the pavement surface material (propagation effect).

Because vehicle mounted systems typically measure noise at the pavement-tire interface, care must be taken to relate on-vehicle measurements to roadside noise levels when the data are to be used for estimating noise impact. Otherwise, the on-board measurements simply give a change between various pavements, not the absolute value of noise that would be experienced at the roadside. Additionally, PFC pavements attenuate sound traveling along the driving lanes and shoulder, a propagation effect that further reduces the roadside noise levels. The OBSI measurements do not capture this effect.

Although use of precision sound meters at roadside provides the best and most relevant measure of traffic noise, it is impractical for use on a large scale due to the setup time and the time required taking the measurements: typically 10–30 min per measurement, three replicates per section. Therefore, on the national and international level, practitioners have migrated toward vehicle-mounted measurement systems to be able to estimate roadside noise levels quickly, at low cost, and with low risk to personnel.

Three classes of on-board systems have emerged, which can be roughly characterized as a) free field close proximity devices, b) enclosed close proximity devices, and c) sound intensity devices. The former two systems are referred to as CPX (generally trailer mounted) and the latter is termed OBSI (vehicle mounted). Much work has been done with all three systems, each having specific advantages and disadvantages, with OBSI currently emerging as the dominant system and the one being used in this project.

The OBSI test method provides an objective measure of the acoustic power per unit area produced as a result of the operation of a vehicle at locations in close proximity to the tire/pavement interface, which allows for the acoustical evaluation of pavements. Dr. Paul Donavan and General Motors (*23*) first developed this near-field measurement method for traffic noise. As the name indicates, the method measures sound intensity, which is defined as the average rate of sound energy transmitted in a specified direction at a point through a unit area normal to this direction at the point. Units are watt per square meter (W/m^2). As such, it is a vector quantity with magnitude and direction, as opposed to sound pressure, which is a scalar quantity. The direction of sound intensity can be associated with the direction of sound propagation or the direction of the orientation of the probe used for measuring sound intensity.

A group of experts from all parts of the United States that have been implementing the method over the last few years has developed an AASHTO Standard (TP 76-13) in an effort to make the procedure uniform, allowing various pavements and textures to be directly compared (*24*). As expert users of this test method, both TxDOT and CTR have been involved in this effort.

The procedure utilizes a fixture positioned close to the tire to hold the sound intensity probe. Details of the equipment utilized are presented in the next section. A single sound intensity probe, such as the one utilized throughout this project, consists of two 25 mm (1 inch) microphones spaced 16 mm (0.6 inch) apart and preamplifiers in a side-by-side configuration. A dual probe consists of two pairs of microphones. A foam windscreen is placed over the microphones to reduce the wind noise. The probe is positioned 100 mm (4 inch) away from the plane of the tire sidewall and 75 mm (3 inch) above the pavement surface, mounted to the rear tire on the passenger side of a car. Signals from the microphones are input into a real-time analyzer. Measurements are taken at 97 kph (60 mph) at two intensity probe locations. One location corresponds to the leading edge and the other to the trailing edge of the tire/pavement contact patch. Figure 5 shows the intensity probe positions and distances in relation to the tire and pavement. A dual probe allows testing at both the leading and the trailing edge positions simultaneously. With a single probe, the tests must be run for one microphone position, and then the microphones must be switched to the other position. Therefore, runs with a single probe take twice as much time.

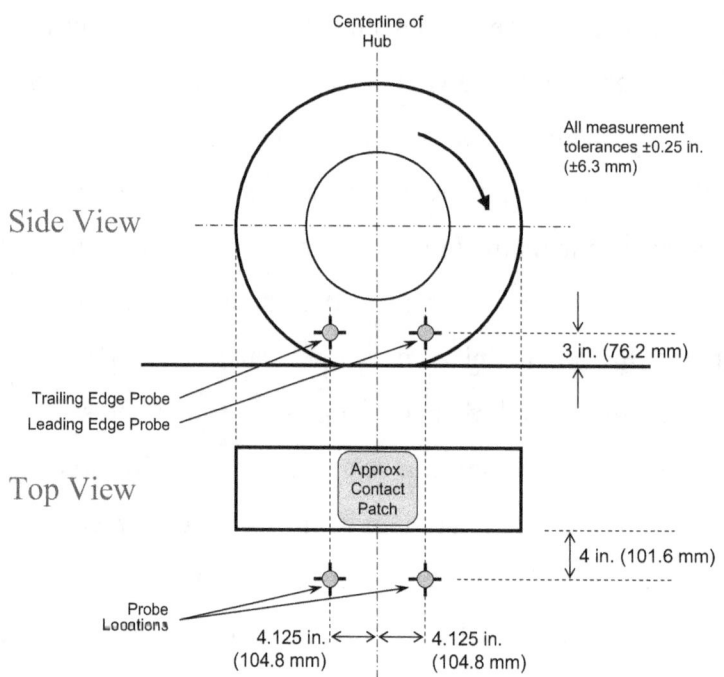

Figure 5. Sound Intensity Probe Showing Leading and Trailing Edges.

At a minimum, two valid test runs shall be performed per probe position, according to the standard. In the majority of the cases in this project, three replicate measurements were collected at each probe location and then averaged to obtain the overall noise levels. Each measurement is averaged over a 5-second period, yielding test sections that, given the traveling speed of 97 kph

(60 mph), are 134 m (440 ft) long. Therefore, a test section is defined as a 134 ± 3 m (440 ± 10 ft) length of pavement over which a sound intensity measurement is made.

The results are reported as overall A-weighted sound intensity levels, and as A-weighted one-third octave band levels. The overall sound intensity level is the sound intensity level corresponding to the energy sum of the A-weighted sound intensity within the one-third octave bands ranging from 400 to 5000 Hz.

$$\text{Overall Sound Intensity Level} = 10 \times \text{Log}_{10}\left(\sum_{i=400}^{i=5000} 10^{(L_i/10)}\right)$$

where L_i is the A-weighted intensity level in the one-third octave band with center frequency i. If a single probe configuration is used, as was the case throughout this project, sound intensity levels from multiple valid test runs are arithmetically averaged for the leading and trailing edges separately. The leading and trailing edge averages then are energy-averaged to calculate a single result that is the average of test runs, commonly referred as the tire average.

The system used to measure the sound intensity using the on-board method is comprised of the following equipment:

- Matched microphone pairs.
- Preamplifiers.
- Cables.
- Sound analyzer.
- Probe holders (fixture)
- Associated items mounted on the test vehicle, the vehicle itself, and the test tires.

In June 2006, the OBSI system, which Dr. Paul Donavan of Illingworth & Rodkin had developed, and which Caltrans used extensively, became available to TxDOT and CTR. CTR obtained the equipment for the development of the predecessor of this project, 0-5185 *Noise Level Adjustments for Highway Pavements in TxDOT* (*10,25,26*). The OBSI fixture is a custom machined jig that bolts to the wheel rim and supports a sound intensity probe at very close proximity to either the front or rear tire/pavement contact point. Because the device is bolted to the wheel, the vertical distance from the pavement does not vary as the suspension oscillates. Since there is a robust bearing connecting the bolted on assembly to the microphone holders, the device does not rotate with the wheels. A slender vertical post affixes to the car body to steady the assembly and provide resistance to the small amount of rotational force generated by friction in the bearing. As mentioned before, the measurements throughout this research project were conducted using a single probe. The first OBSI fixture can be seen attached to one of the test vehicles (see Figure 6). In this fixture, the microphones are in a horizontal position, parallel to the ground.

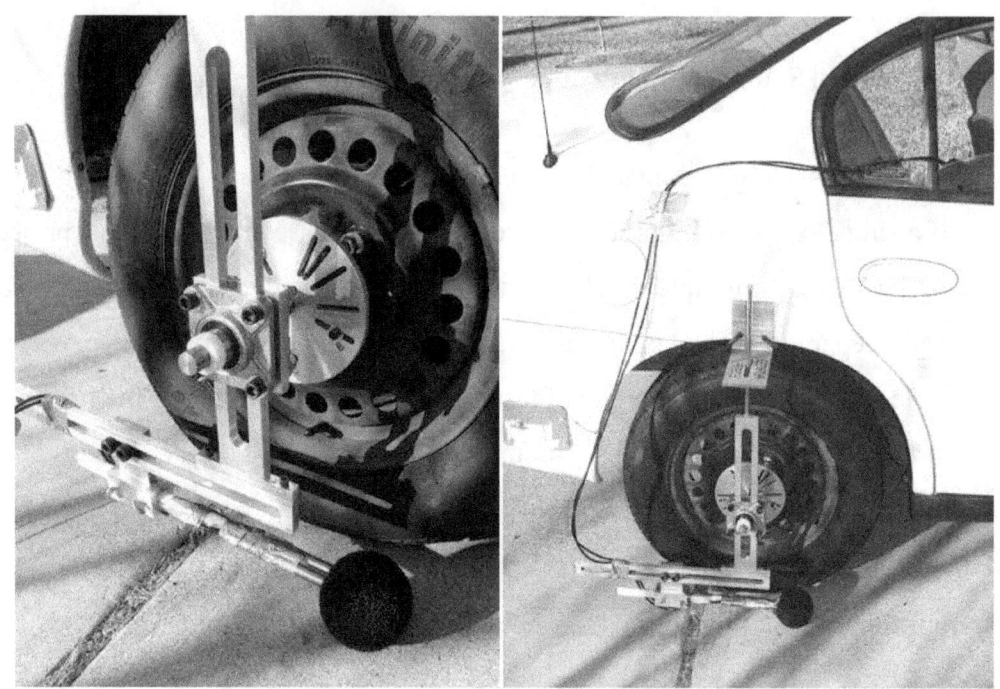

Figure 6. Single Probe Fixture with Microphone Pair in Horizontal Position.

In 2010, CTR and TxDOT started using a new fixture, featuring an improved design that allowed for better stability of the microphone pairs and reduced vibration. Although this new fixture is designed as a dual probe, the fixture could only be utilized as a single-probe because the analyzer used for the measurements is a two-channel device. This means that the microphone pair must still be switched between leading and trailing edges halfway through the runs to complete the measurements. The new fixture holds the microphone pairs in a vertical position (see Figure 7).

The system utilizes two 12.5 mm (0.5 inch) phase-matched condenser free-field microphones (see Figure 8). Preamplifiers are affixed to each individual microphone for signal amplification, and these, in turn, are attached to a plastic probe holder that keeps a space of 16 mm (0.63 inch) between microphones, in a side-by-side configuration. The microphones are fitted with a spherical windscreen.

G.R.A.S. manufactured the microphones and preamplifiers utilized for this project. These items comply with the requirements of the international standard IEC 1094 for Measurement Microphones, and as required by the AASHTO OBSI Standard, and also comply with the Class 1 requirements of ANSI S1.9. These devices are able to measure the real part of a complex sound intensity in sound fields with a high level of background noise, such as occurs on the highway.

Figure 7. New OBSI Fixture.

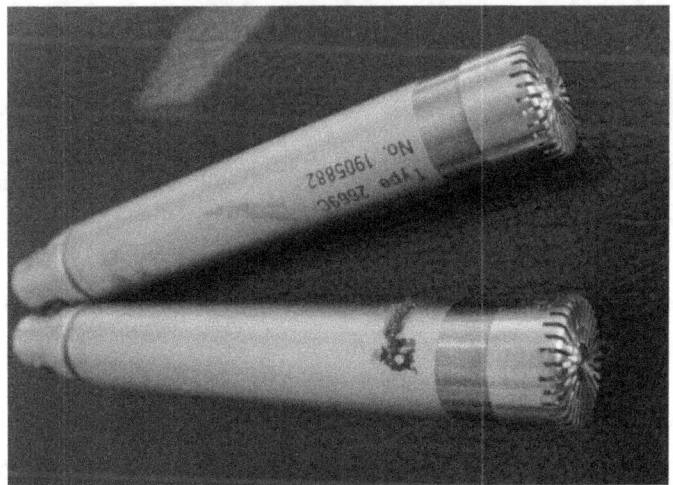

Figure 8. Half-Inch Sound Intensity Microphone Pair.

OBSI data are not only collected in raw form, but also analyzed in real time using an in-vehicle device, the Larson Davis 3000+ Analyzer (see Figure 9). The analyzer affords the advantage of analyzing and displaying the sound intensity for each frequency band in real time, allows the operator to listen to the noise being recorded, and stores the processed data inside the analyzer itself. A separate, flash card audio recorder captures the raw output from the LD 3000 for reprocessing later, if needed.

Seeing the data display and hearing the noise during the test is extremely useful, as it allows the operator to immediately detect any anomalies during testing, and flag the data as suspicious for later examination. Often, mechanical problems with the jig or vehicle occur that the operator can easily detect and correct in the field, so that testing can resume.

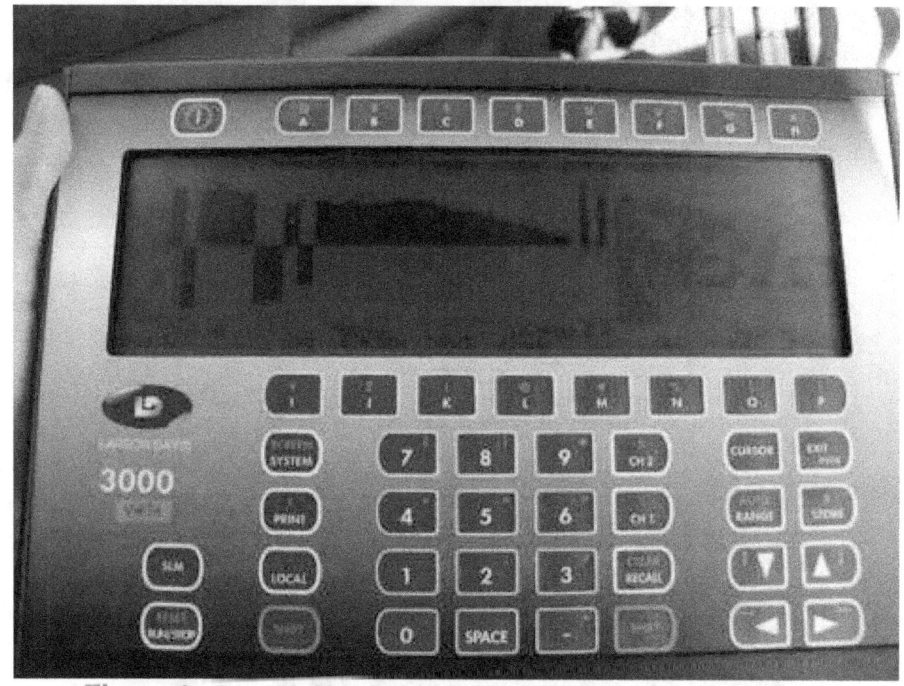

Figure 9. Larson Davis 3000+ Sound Intensity Analyzer.

In an experiment to evaluate the amount of engine noise that the OBSI device had measured, the research team performed a set of runs on FM 620 in Austin, in which the vehicle remained parked while the engine RPMs were raised to an equivalent level to that of the OBSI method testing speed (97 kph [60 mph]). The noise measured with the stationary test was then compared to a run on the pavement section at 97 kph (60 mph) demonstrating that the engine noise contribution to the measured level was insignificant compared to the overall tire/pavement noise.

The main disadvantage of the LD 3000 analyzer is that it has input for only two channels, meaning that even with a dual probe fixture, only one pair of microphones could be used at a time. Because of this, as mentioned before, the microphones still had to be switched between the two testing positions of leading and trailing edges. A four-channel system would have been necessary to run leading and trailing edges simultaneously.

The test tires and the test vehicle are other fundamental components of the system. The standard only specifies that the test vehicle should be a passenger car, in which the test tire is not covered on the outboard side. The load on the test tire due to the weight of the vehicle including passengers, test hardware, fuel, and other contents shall be 360 ± 45 kg (800 ± 100 lb) during the test, according to the standard.

Two types of vehicles were utilized throughout this project: a Chevrolet Malibu and a Chevrolet Impala. The use of two vehicle types is directly related to the test tires. When the project began, CTR had the Chevy Malibu vehicle (Figure 10) equipped with Uniroyal Tiger Paw All-Weather Performance (AWP) tires. The AWP is a P205/50R15 tire. This type of tire and vehicle were common among the various options that OBSI practitioners used at the time, when the AASHTO standard started to be drafted. At this time, there was no specification for tires, so there were three or four common types of tires that were normally selected for OBSI tests. Testing for this project took place with that vehicle-tire combination until 2010.

Later on, the AWP tire stopped being manufactured, and it became increasingly difficult to find commercially. In order to transition to a more available tire and to devise a uniform test procedure capable of producing comparable results among different agencies utilizing the OBSI method, the expert panel group drafting the standard decided to adopt the ATSM P225/60R16 Standard Reference Test Tire (SRTT) as the standard tire for OBSI tests. Therefore, halfway through the project the research team had to proceed with new tires and a new vehicle. The SRTT does not fit in the Malibu vehicle, so for the remaining part of the project a Chevy Impala equipped with SRTTs had to be utilized (see Figure 11).

The standard specifies that the test tire shall be mounted on the vehicle on a non-driven, non-steering axle or hub (i.e., the test tire shall not be a wheel directly driven by the powertrain or steering assembly). The tire shall be mounted so that rotation is in the same direction for the life of the tire. The standard for the tire is ASTM F 2493, Standard Specification for P225/60R16 97S Radial Standard Reference Test Tire (*27*). The change in tires that occurred during this project had a definitive influence on the test results, as the tire size and tread pattern are substantially different (see Figure 12).

Figure 10. Chevy Malibu Test Vehicle.

Figure 11. Chevy Impala Test Vehicle.

Figure 12. SRTT Tire (top) and AWP Tire (bottom).

Drainability Measurements

One of the main functions of a PFC pavement is to drain water quickly from the surface. WFV is a measure of how quickly a known volume of water will drain through an area of the PFC pavement surface, and serves as an indicator of how the pavement will perform during a rain event. Based on Tex-246-F, the apparatus used to measure WFV is a 150 mm (6 inch) diameter by 450 mm (18 inch) tall permeameter made of PVC with a clear pipette attached to the side that allows the user to see the water level in the permeameter (see Figure 13). Plumber's putty is used around the base of the permeameter to create a watertight seal between the permeameter and the pavement. It is important that no water can seep through the putty at the base of the permeameter since that would yield a false indication of the WFV of the pavement.

To perform the test, water is poured into the permeameter to a few inches above a marking line on the pipette. Figure 14 shows the permeameter setup at the start of the test when water is poured in. When the water level gets to the top start line, timing begins. The test concludes when the water level has reached a second line located 10 inches below the start line on the pipette. This stop line is 114 mm (4.5 inch) above the pavement surface. The time it takes for the water level to go from the start line to the stop line is the recorded WFV. The test is performed both OWP and BWP and occasionally on the shoulder of the road. At the time of construction, an acceptable WFV is less than 20 seconds (*28*). Subsequent measurements can range from under 20 seconds to several minutes. For practical purposes, a cut off of 5 minutes was established in the field to terminate the measurement and label the pavement as impervious.

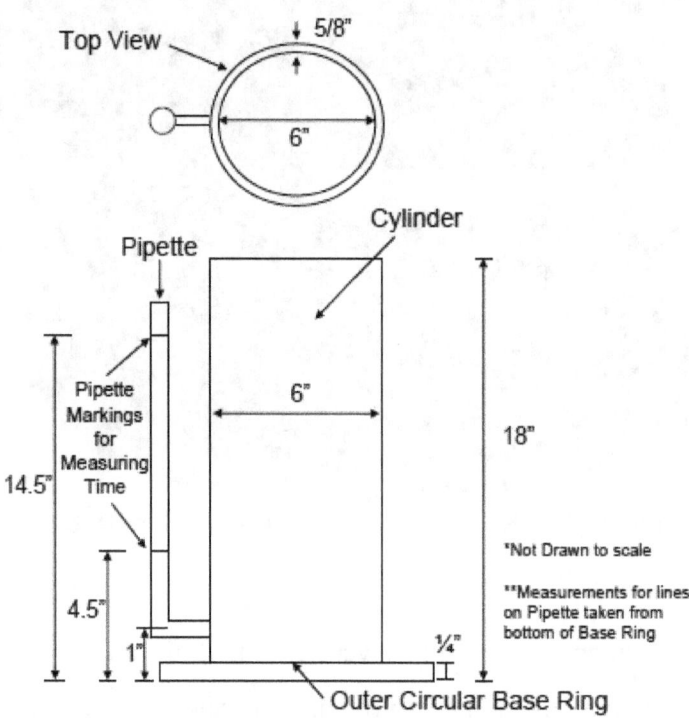

Figure 13. WFV Permeameter Dimensions (*28*).

The WFV test offers a quick and simple way of quantifying drainability (not to be confused with permeability) in PFC pavements. Ideally, a PFC will drain rainwater quickly from the surface, which will reduce the risk of hydroplaning and reduce the amount of splash and spray experienced by motorists. The drainability comes from the relatively high volume of interconnected AV that is introduced into the pavement. Typically, a design will call for at least 18 percent AV at the time of construction, although consolidation from vehicle traffic and clogging of voids will reduce this value over time.

Figure 14. WFV On-Site Test Setup.

It is useful to compare the WFV OWP vs. BWP to get a sense of just how much the traffic is contributing to the consolidation of the PFC. With consolidation the drainability of the PFC will naturally decrease, resulting in higher WFVs over time and decreased functionality. By comparing the WFV of a PFC over several years, it is possible to track this aspect of pavement performance.

Based on findings from TxDOT Project 0-5262, the standard deviation for WFVs at a single location is measured as 0.36 seconds. However, the WFVs between several different locations on a PFC site vary considerably more. These differences can be attributed to different loading patterns, uneven compaction at the time of construction, and varying degrees of clogging within the PFC structure. Some PFC pavements may become so compacted that these become practically impervious, in which case the PFC offers little advantage over typical HMA pavements with regard to drainability.

Durability Measurements

Due to a higher-than-normal binder and AV content, PFCs are more susceptible to bleeding, rutting, or raveling. Repeated traffic loads may also cause excessive compaction leading to closing up of the open pore network in PFCs. In addition, accumulation of dust and debris tends to clogs the AV structure, limiting the pavement drainability.

In order to evaluate the integrity of the pavement, a close-up visual inspection was performed at each site to detect raveling, rutting, clogging, and other pavement defects. A level and ruler were used to measure the depth of rutting in both wheel paths when rutting was apparent. Pictures were taken at each location for later reference and documentation.

When severe pavement defects were encountered, especially raveling, several core specimens were acquired for forensic investigation. These cores were used to determine pavement layer thickness and were further tested in the laboratory for permeability, AV content, aggregate gradation, and binder content.

Texture Measurements

The two most widely used methods for measuring mean profile depth (MPD) are the sand-patch method and the CTM. The sand-patch method uses a known volume of either uniformly small glass beads or Ottawa sand (which is also spherical) to determine the MPD of a pavement. The material is poured onto the pavement and then it is carefully spread out in an approximately circular area until the material evenly fills all depressions between and around aggregates on the pavement surface. The uniformly smooth area of the sand (or glass bead) circle is then measured to determine the diameter from several angles. The area of the sand-patch is used in conjunction with the known volume of the material to calculate the MPD of the pavement. The sand-patch test is relatively quick and simple to perform, although limited experience can result in inaccurate results. PFC pavements are designed to have a porous structure and because of this, the sand-patch test cannot be performed on PFC pavements. The interconnected pores in PFC allow sand to disperse to subsurface pores. This will cause the area of a sand-patch to be smaller than expected and will result in an MPD that is higher than the real value.

NCHRP Synthesis 291 provides a summary of survey findings from the United States and foreign transportation agencies (*29*). Based on this report, no agency in the United States provides minimum texture or friction coefficient values. However, in Great Britain a minimum mean texture depth (MTD) of 1.5 mm (0.059 inch) is recommended for new pavements. The report also summarizes survey responses for threshold texture values prior to maintenance and rehabilitation. In Quebec, the minimum value for MTD is 0.6 mm (0.023 inch). A study conducted in New Zealand recommends a minimum MPD of 0.7 mm (0.027 inch) for posted speeds greater than or equal to 70 kph and a minimum MPD of 0.5 mm (0.020 inch) for speeds less than 70 kph (44 mph) (*30*).

The equipment used to measure MPD on PFC pavements is the CTM that Nippo Sangyo Company of Japan developed. The ASTM Standard E2157-09 describes the proper use of the CTM. This machine has a laser displacement sensor that rotates on an arm for a specific circumference and records the measurement of the texture approximately every 0.9 mm (0.035 inch). On other wearing surface pavement layers besides PFCs, the sand-patch method and the CTM are shown to correlate well. Figure 15 shows the apparatus used in the field.

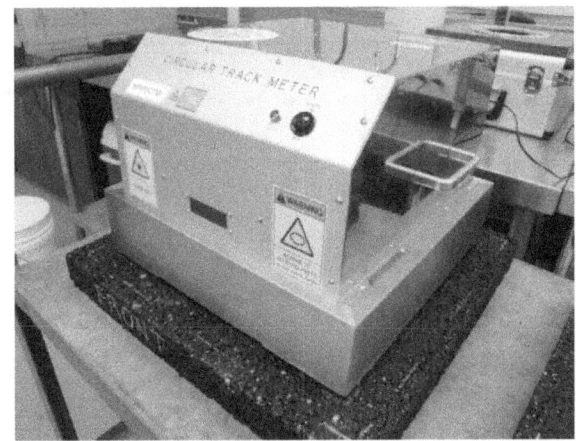

a) Side View   b) Bottom View

Figure 15. CTM Apparatus.

Friction Measurements

In addition to texture, wet pavement friction is another important variable used to assess the performance of PFC pavements in terms of safety. Both microtexture and macrotexture influence the pavement friction coefficient with microtexture playing a larger role at lower speeds. As speed increases, the macrotexture gradually overtakes microtexture in the influence over friction coefficient. Additionally, as speed increases, pavement friction tends to decrease (see Figure 16), where the x-axis shows the speed in kph and the y-axis is the coefficient of friction.

The calculation for pavement friction is relatively straightforward. As shown in the simple diagram of Figure 17, when a rubber tire skids on the pavement from braking action, the weight of a portion of the vehicle (W) causes an equal and opposite normal force (N) with the pavement. The momentum of the moving vehicle causes a force (P) in the direction of travel, which is opposed by a force of friction (F), proportionally related to N by the friction coefficient μ. The friction coefficient μ is constant at a constant speed, and it will increase or decrease with a change in speed depending on the properties of the two materials in contact, the tire and the asphalt pavement surface in this case. The equation of friction force is F=μN, which can be rearranged to μ=F/N.

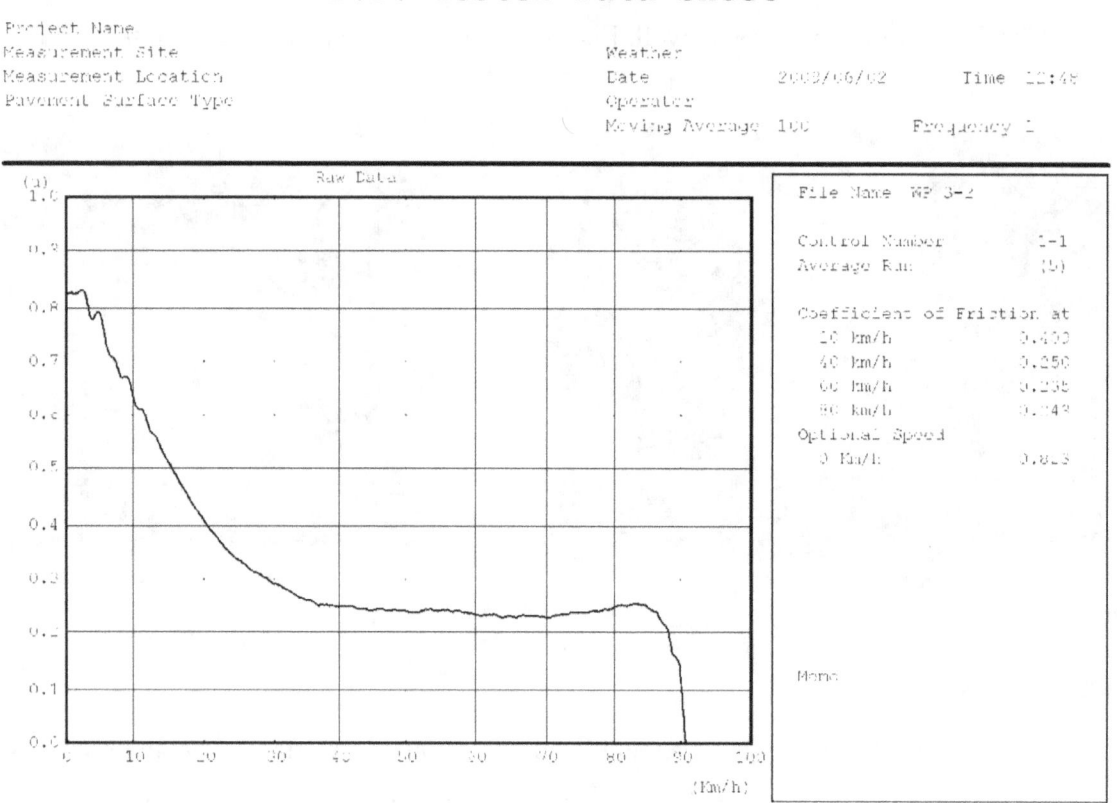

Figure 16. Typical DFT Output.

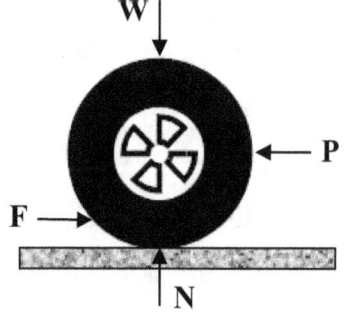

Figure 17. Tire-Pavement Friction Force Diagram.

The DFT is a portable tool for measuring the dynamic friction coefficient. Figure 18 shows the equipment used for on-site testing. In addition to the apparatus, a steel water bucket connects to a nozzle on the DFT through a plastic hose. The bucket provides gravity-fed water to the DFT, which is then discharged through the bottom of the unit during testing to wet the pavement. A portable processing unit and laptop computer connects to the DFT to control and monitor the apparatus and collect the test results.

The DFT has three rubber sliders that are spring-mounted to the bottom of a disk with a diameter of 350 mm (14 inch) (see Figure 18b). The disk is held above the pavement as the electric motor in the DFT spins the disk so that the tangential velocity of the sliders reaches 90 kph (56 mph). The DFT sprays water onto the pavement. Upon reaching the target testing velocity, the motor disengages and the rubber sliders are lowered onto the pavement surface. After contact with the pavement surface, the disk maintains a constant load and begins to decelerate. A transducer on the disk measures the friction force and accompanying speed of the disk. From this data, a continuous spectrum of the resulting friction coefficient is generated from 90 to 0 kph (56 to 0 mph) as previously shown in Figure 16. Of this spectrum, the friction coefficients at 20, 40, 60, and 80 kph (12, 25, 37, and 50 mph) are recorded into the 0-5836 project database.

a) Side View   b) Bottom View
Figure 18. DFT Apparatus.

Skid Measurements

Skid resistance is an important characteristic useful in determining when to schedule pavement maintenance. With time, roads lose skid resistance due to vehicles wearing down the pavement surface as well as depositing oils. Additionally, excessive bleeding can further reduce the skid resistance of the pavement. To quantify the skid resistance of roads in Texas, TxDOT began collecting skid resistance data in 1973. Each year TxDOT collects these data on 25 percent of the state highway system and 50 percent of the interstate highway system. Data collected are stored in PMIS and uploaded to TxDOT mainframe for TxDOT's engineers to use in various pavements analyses.

Currently, TxDOT does not specify a minimum skid number (SN) threshold for PFCs or other pavement surfaces. According to Jayawickrama et al. (*31*), some states usually consider SNs of 30 and above to be acceptable for low-volume roads and 35–38 as acceptable for high-volume roads. When the SN falls between 30 and 34, states recommend frequent monitoring of the pavement; if it falls below 30, corrective actions are recommended to improve skid.

TxDOT conducts skid testing using a two-wheeled trailer that is towed behind a truck (see Figure 19). The left tire of the trailer locks during testing and an electronic torque ring strain gauge located on the axle of the skidding tire measures strain that is further recorded on an on-board computer. During testing, a pressurized water sprayer wets the road in front of the locking tire. The water serves to lubricate the road to avoid causing the trailer to jerk to one side. In addition, the water serves to simulate skid resistance during wet conditions.

Figure 19. Two-Wheeled Trailer with Left Wheel Path Locking Tire.

Measurements are taken each half-mile to minimize wear on the skid tire. The steps performed during an individual skid cycle are as follows:

1. Spray water in front of tire.
2. Lock the wheel ½ second after the water begins to spray.
3. Skid the wheel along the pavement.
4. After skidding 23 m (75 ft), release the brake.
5. Stop water spray ½ second after tire is unlocked.

Figure 20 shows an example of the data collected after a single cycle. The x-axis is time in hundredths of seconds and the y-axis is the skid resistance as calculated from the torque resistance. The SN is calculated from the data in the yellow area of the figure and the red line shows the SN. The blue box labeled GPM marks the water flow in gallons per minute.

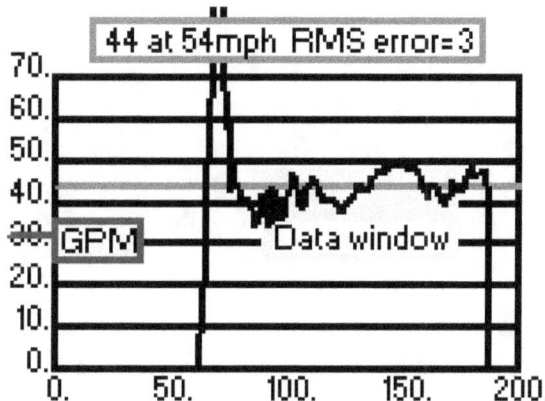

Figure 20. Typical Graph Output Obtained after a Skid Cycle.

Accident Data Collection

Aside from monitoring the performance properties of the selected pavement sections, it was also important to evaluate if PFCs were improving safety by reducing the number of accidents, especially under wet weather conditions. TxDOT is responsible for the collection and analysis of all reported motor vehicle traffic crashes that law enforcement offices throughout Texas have submitted. For this purpose, TxDOT maintains a Crash Records Information System (CRIS) database covering crash records from 2003 to 2011. CRIS provides the date and location of each accident along with relevant information pertaining to the crash such as road alignment and condition as well as lighting and weather conditions. The locations of accidents are reported in geographical coordinates and in displacements from Texas reference marker locations. The latter coordinate system was referenced against the beginning and ending reference markers of the pavement sections included in this project.

CHAPTER 4:
PAVEMENT PERFORMANCE

This chapter discusses the results of the on-site data collection performed from 2009–2012. The functionality of the pavement sections is described first with details from the noise and WFV testing. The safety aspect of the pavement sections is described next by texture, friction, skid, and accident data. For WFV, texture, and friction, a comparison between OWP vs. BWP measurements is presented first, followed by an evaluation of the change in each variable with time, and the effect of mixture/binder type, aggregate classification, and climate. The effect of time, mixture/binder type, and climate is also analyzed for the noise data. Appendix D includes detailed noise results for each pavement section.

NOISE

Noise data were collected with the OBSI method on 21 PFC and dense-graded HMA reference pavement sections since 2009. Prior to that, a predecessor project, TxDOT Project 0-5185, collected sound intensity data on all pavement types for sections all over the state (*10,25,26*). Before 2011, all of CTR's noise data collection efforts were conducted using the AWP tire. The new AASHTO standard for OBSI tests includes the utilization of a standard tire, known as the SRTT. In 2011, CTR purchased SRTTs for the test vehicle, and all the project sections were measured with the SRTT. From then on, testing of the project sections was performed using this tire.

Given the same pavement and ambient conditions, the SRTT produces higher overall sound levels compared to those obtained with the AWP tire. This is widely known and documented in the literature (*10,32*). However, since the data collected with the AWP tire were of significant value to this research, it was necessary to develop a correlation to update the older noise values measured with the AWP tire to make them comparable to the SRTT results. Two correlation models were developed using noise measurements acquired over the years as part of this project and as part of TxDOT Project 0-5185.

Existing Data for Correlation

Several noise rodeos have been conducted with TxDOT and in some instances, with other agencies as well, in which the same pavement sections are run using different systems at the same time, for comparison purposes. Some of these rodeos took place as part of TxDOT Project 0-5185, and were very useful in enabling a comparison of TxDOT's and CTR's OBSI systems. Other rodeos were conducted as part of this project. The instances in which the noise rodeo data are useful for establishing the correlation are those in which TxDOT ran its system using the SRTT while CTR ran the AWP tire.

Prior to TxDOT's acquisition of their SRTT, the rodeos confirmed that the TxDOT system and the CTR system were almost identical in their results. The first rodeo took place in September 2007. The mean difference between the CTR and TxDOT vehicles using the AWP tire was 0.6 dBA, with a standard error of 0.16, giving a 95 percent confidence interval ranging from −0.28 dBA to −0.88 dBA, an insignificant difference. Therefore, it was established that measurements taken using the two vehicles could safely be used interchangeably. There was a 97 percent correlation for the data between the two systems. At this stage, TxDOT's and CTR's equipment were considered equivalent.

Correlation Model

In early 2008, a second rodeo was conducted with TxDOT, but this time TxDOT incorporated the use of SRTTs into their system, while CTR kept using the AWP tires. Both agencies were still using the Chevy Malibu as test vehicle. TxDOT made some modifications to its vehicle to accommodate the installation of the larger SRTT in the rear axle. A third rodeo was conducted in July 2008. ANOVA was used to produce such correlation model, in which every other variable was controlled except for the tire type. A 94 percent correlation was obtained in this model, which was developed with the data from the second and third rodeos (see Figure 21).

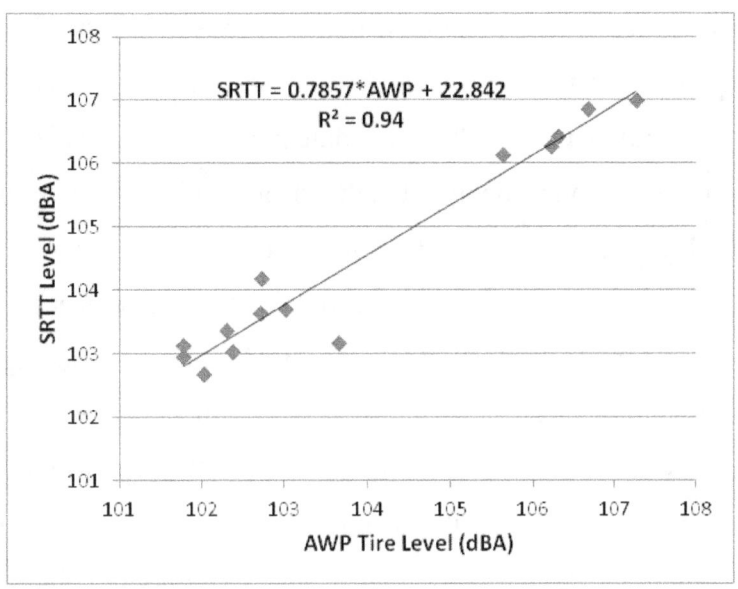

Figure 21. AWP-SRTT Correlation Using the Same Test Vehicle.

The correlation model was developed with data collected on different pavement types, with the same vehicle type (Chevy Malibu), utilizing a brand-new SRTT, and a fairly new AWP tire. The data points that were collected in the measurements indicate that the SRTT levels were higher than their corresponding AWP tire values. This fact confirms that the SRTT tire is slightly louder than the AWP tire, as is commonly accepted. This model was used to make the AWP tire measurements comparable to the SRTT measurements.

Results

Results of the noise measurements are presented next, including graphs of the OBSI analysis that show the overall sound levels over the years throughout this project, and further back in time if available. The overall levels that were originally measured with the AWP tires were adjusted with the correlation model to enable the comparison with SRTT results. In general, the charts show that the noise reached its highest levels with the 2010 measurements, and then dropped. The reason might be that around that time, the aging AWP test tire that was being utilized for the noise tests reached its peak in hardness, before a new SRTT replaced it for the 2011 tests. The 2012 measurements, in general, represent another decrease in levels, as the tests for this period were conducted with an even newer SRTT. As tires get older, the rubber gets harder because of the elements and wear, so the measured noise levels tend to be higher.

The general, most accepted hypothesis is that the acoustic benefits that PFC pavements provide tend to diminish over time, since the physical characteristic that gives these pavements its noise absorption capability—namely the void content of the surface layer—is reduced as the pavement wears down. The rationale behind it is that debris will clog the voids, and traffic loads will compact the PFC layer, resulting in louder noise levels as the PFC ages.

In order to analyze the noise measurements obtained in this project from the pavement age standpoint, the OBSI overall sound levels were plotted for each section against the age of the section at the time of each particular test. Age was computed in years, from the time of construction for each pavement to the date when the OBSI test was performed. The results are shown in Figure 22, which includes both PFCs and dense-graded HMA reference sections.

As shown in Figure 22, there was not a clear pattern of noise level trend with age; even some sections showed that these got quieter with time. A possible explanation for the reduction of noise levels with age is that, as indicated before, the last set of measurements (summer 2012) was performed with a brand-new tire, whereas the 2011 tests were conducted with an almost-new tire, and the measurements prior to 2011 were conducted with older AWP tires.

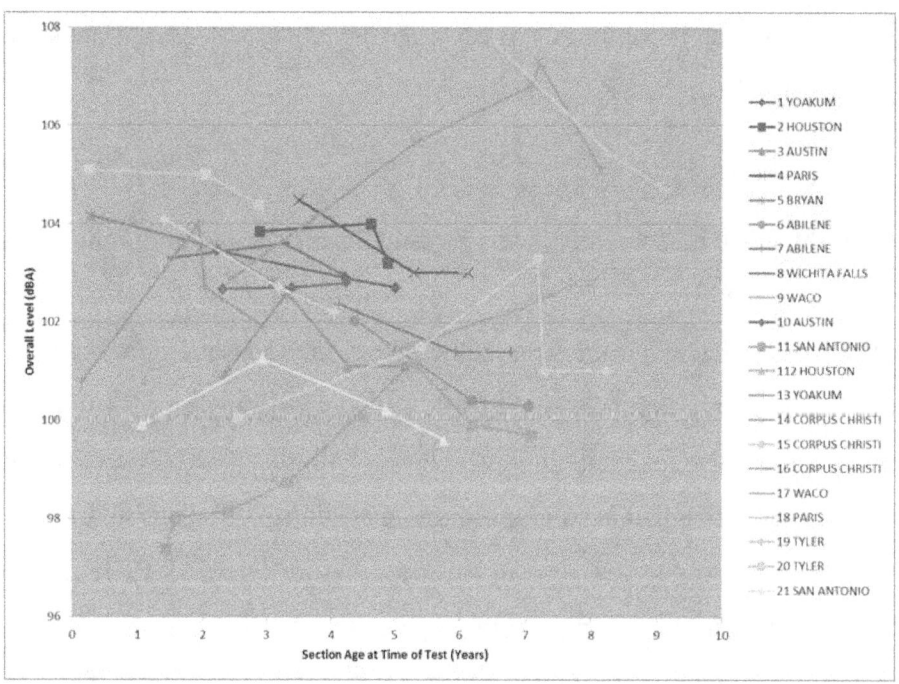

Figure 22. OBSI Noise Levels with Pavement Life.

As tires age, the rubber hardens, resulting in higher noise levels. Therefore, a brand-new tire is expected to provide lower noise levels. The correlation model was applied to noise measurements performed prior to 2011 (measured with AWP tires) in an effort to make the results comparable with tests conducted with SRTTs. However, no adjustment was made to account for tire age.

If the overall levels are plotted against pavement age as shown in Figure 23, disregarding the association of each measurement with its pavement section, it can be seen that there is a slight upward trend in noise levels with pavement age. The correlation between noise levels and age, however, is rather poor ($R^2=0.02$).

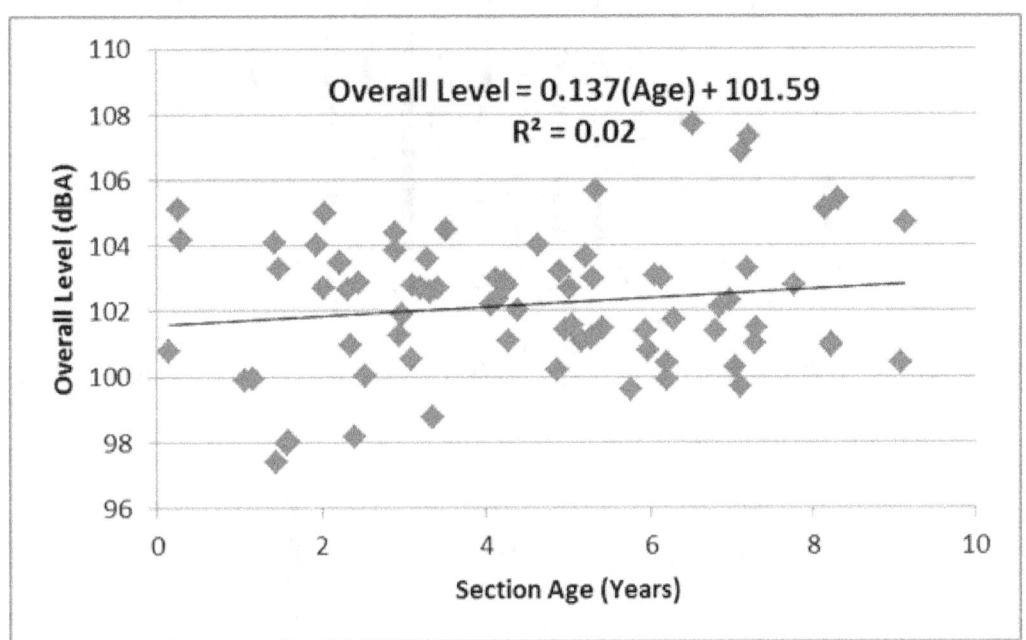

Figure 23. Correlation between Noise Levels and Pavement Age.

The effect of binder/mixture type on noise levels was also evaluated. Among the pavement sections studied for noise, there are 16 PFCs and five dense-graded HMAs designated as reference pavements. From those 16 PFC pavement sections, five include AR binder and the other 11 include PG 76-22 binder. Previous research on PFCs mixes in Texas, performed under TxDOT Project 0-5185 (*10*), showed that AR binder mixes result in quieter PFCs compared to those constructed with PG binder.

Graphs for the overall levels for each section corresponding to 2009, 2011, and 2012, sorted by pavement type and showing the binder type are presented in Figure 24, Figure 25, and

Figure 26, respectively. Figure 24, Figure 25, and Figure 26 also display the variability of the measurements for each of the sections in the form of standard deviation bars. The chart for the 2010 measurements is omitted because only a few sections were tested that year.

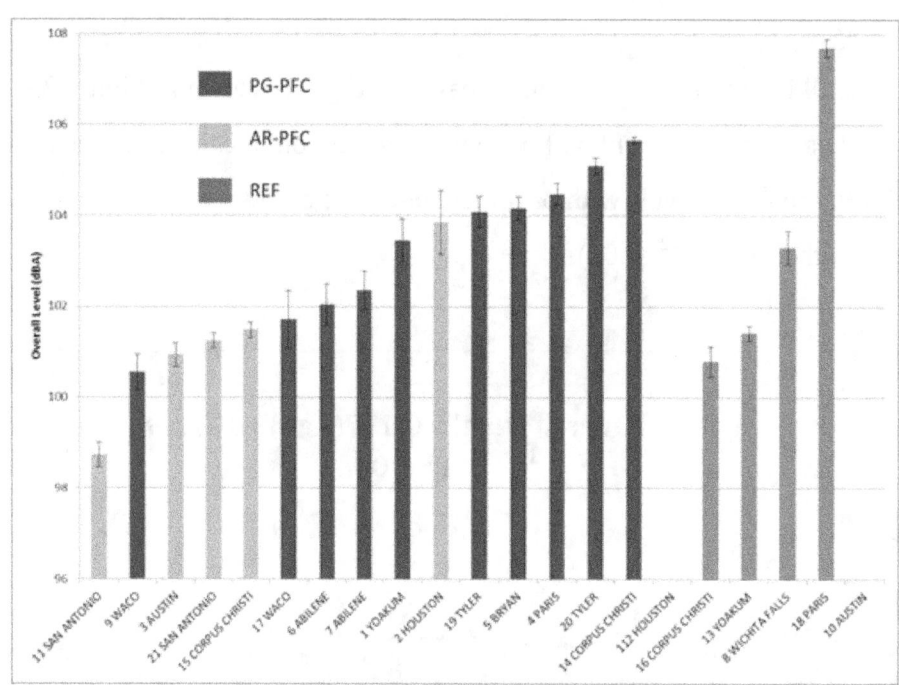

Figure 24. Effect of Binder/Mixture Type on Noise Levels for Measurement Year 2009.

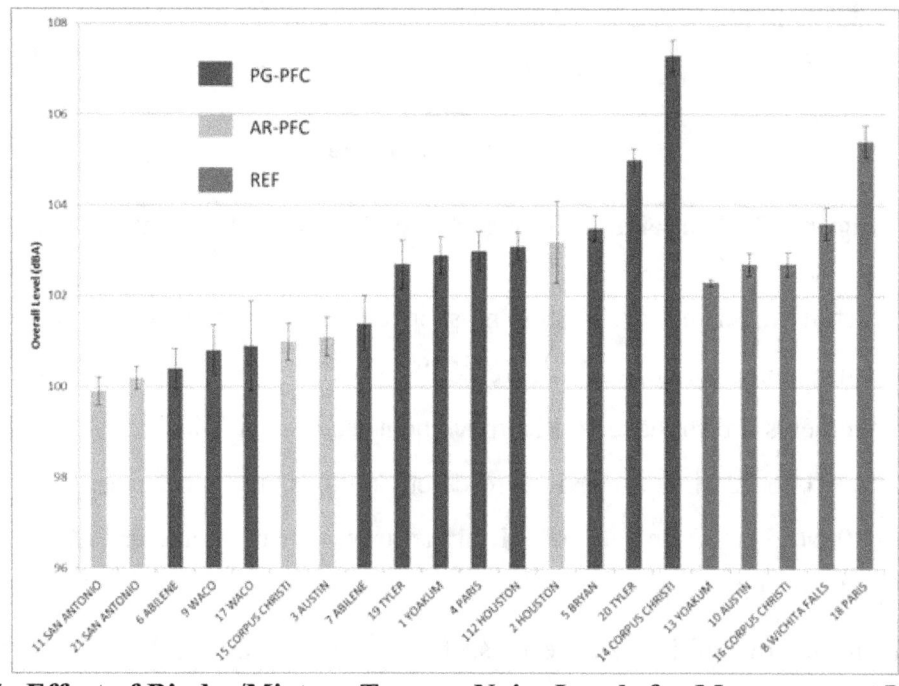

Figure 25. Effect of Binder/Mixture Type on Noise Levels for Measurement Year 2010.

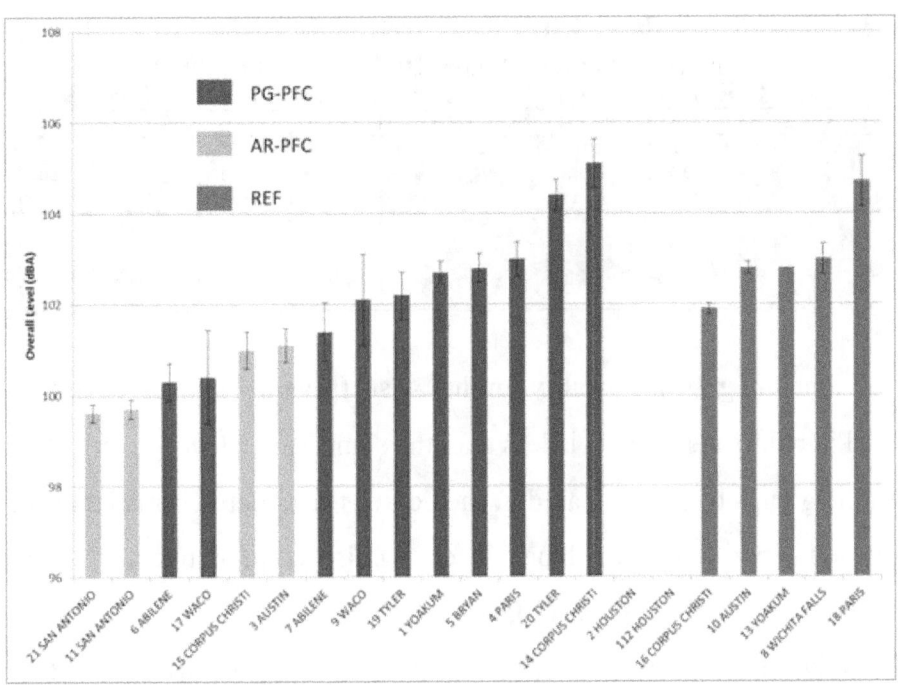

Figure 26. Effect of Binder/Mixture Type on Noise Levels for Measurement Year 2012.

Figure 24, Figure 25, and Figure 26 show that, in general, AR-PFCs appear to be quieter than the PG-PFCs, and that PFCs in general are quieter than dense-graded HMAs, although the differences are small, but the trends are consistent throughout the years. Unfortunately, the number of AR-PFC sections is very small and thus drawing more definitive conclusions in this regard is difficult. Another important aspect that can be observed is that not all dense-graded HMA reference sections are louder than PFCs. In addition, a few sections are consistently quiet (e.g., Sections 11 and 21, in San Antonio) and some are consistently loud (e.g., Sections 20 in Tyler, and 14 in Corpus Christi, both PFCs, and Section 18 in Paris, a reference section).

To assess the significance of the differences in noise levels measured on PFCs vs. the reference sections, and the significance of the differences in noise levels measured on AR-PFCs vs. PG-PFCs, the research team conducted t-tests. For each case, a two-sample t-test was performed; the assumptions were a) equal variances, $\alpha=0.05$, and b) a two-tailed distribution. These tests were done for the 2009, 2011, and 2012 sets of data. Table 2 shows the results.

Table 2. T-Test Results for the OBSI Noise Levels with Respect to Binder/Mixture Type for Measurement Years 2009, 2011, and 2012.

	2009 PFC vs. REF		2009 AR-PFC vs. PG-PFC		2011 PFC vs. REF		2011 AR-PFC vs. PG-PFC		2012 PFC vs. REF		2012 AR-PFC vs. PG-PFC	
	PFC	REF	AR	PG	PFC	REF	AR	PG	PFC	REF	AR	PG
Mean (dBA)	102.7	103.3	101.3	103.4	102.3	103.3	101.1	102.8	101.8	103.0	100.4	102.4
Variance	3.717	9.745	3.300	2.673	3.874	1.553	1.667	4.106	2.760	1.043	0.657	2.380
Observations	15	4	5	10	16	5	5	11	14	5	4	10
Pooled Variance	4.781		2.866		3.385		3.409		2.356		1.950	
t Stat	-0.519		-2.271		-1.130		-1.745		-1.497		-2.530	
P(T<=t)	0.610		0.041		0.273		0.103		0.153		0.026	
t Critical	2.110		2.160		2.093		2.145		2.110		2.179	

The row containing the probability for the t-tests (p-value) is the second to last, and it has been highlighted in red. If the p-value is less than the significance level α, the difference in noise levels between the groups being compared in each case is considered statistically significant. This only happened in two cases, which have been highlighted in orange: the 2009 AR-PFC vs. PG-PFC, and the 2012 AR-PFC vs. PG-PFC.

For the 2009 set of measurements, the tests indicate that the difference between PFC and reference sections was not statistically significant, but the difference between AR and PG for PFCs was significant. The 2011 results show no statistically significant differences between PFCs and reference sections, and no statistically significant differences between AR-PFCs and PG-PFCs either. For the case of the 2012 measurements, the t-test results show that the noise levels for PFCs and reference sections were statistically equivalent, while a statistical significant difference was detected between AR-PFCs and PG-PFCs.

For the comparisons between PFCs and reference sections, it was expected that the result would show significant differences, because PFCs and dense-graded HMA pavement sections are expected to have acoustical performance differences. However, that was not what the t-tests indicated. For the comparison between binder types, both the 2009 and 2012 measurements indicated that there was a significant difference between AR and PG, among the PFCs.

The data were also analyzed as a single dataset, i.e., with all the noise level measurements together, regardless of the measurement year. The overall comparisons between PFC and reference sections and between AR-PFC and PG-PFC were evaluated in a similar manner, with t-tests. Table 3 shows the results of these tests. Even though there is a difference in the mean noise levels between PFCs and Reference sections showing that PFCs were quieter, such a difference is not-statistically significant. For the case of AR-PFC and PG-PFC, the difference in means indicating that AR binder yields quieter mixes than PG binder was statistically significant, as shown in the result highlighted in orange in the table.

Other observations from the statistical analysis are that PFCs had higher variability than the reference sections, and among PFCs, those with AR binder had less variability than those with PG binder.

Table 3. T-Test Results for the OBSI Noise Levels with Respect to Binder/Mixture Type for All Measurement Years.

	PFC vs. REF		AR-PFC vs. PG-PFC	
	PFC	REF	AR	PG
Mean	102.4	103.2	101.4	102.9
Variance	3.391	2.720	2.159	3.255
Observations	54	16	19	35
Pooled Variance	3.243		2.876	
t Stat	-1.668		-3.239	
P(T<=t)	0.100		0.002	
t Critical	1.995		2.007	

The effect of climate on noise levels was explored based on the climatic regions within the state of Texas (Figure 3): dry cold (DC), dry warm (DW), moderate (M), wet cold (WC), and wet warm (WW). Table 4 lists the classification of the sections where noise was measured. Graphs presenting the noise measurements for each of the sections, grouped by climatic region and sorted by noise level within each region, are shown for the 2009, 2011, and 2012 in Figure 27, Figure 28, and Figure 29, respectively.

Table 4. Classification per Climatic Region of Pavement Sections Used for Noise Measurements

DC	M	WW	DW	WC
6 ABILENE	3 AUSTIN	1 YOAKUM	11 SAN ANTONIO	4 PARIS
7 ABILENE	9 WACO	2 HOUSTON	14 CORPUS CHRISTI	18 PARIS
8 WICHITA FALLS	10 AUSTIN	5 BRYAN	15 CORPUS CHRISTI	19 TYLER
	17 WACO	112 HOUSTON	16 CORPUS CHRISTI	20 TYLER
		13 YOAKUM	21 SAN ANTONIO	

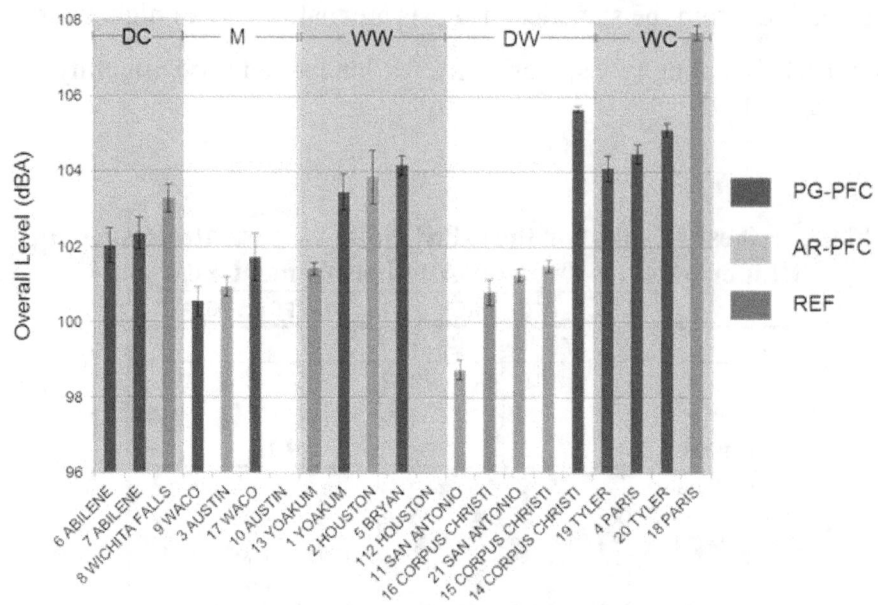

Figure 27. Effect of Climate on Noise Levels for Measurement Year 2009.

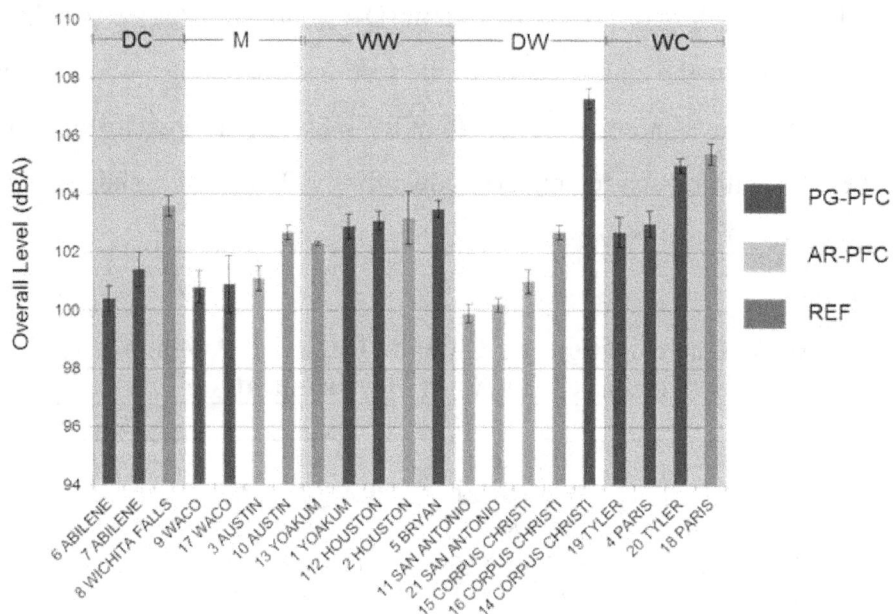

Figure 28. Effect of Climate on Noise Levels for Measurement Year 2011.

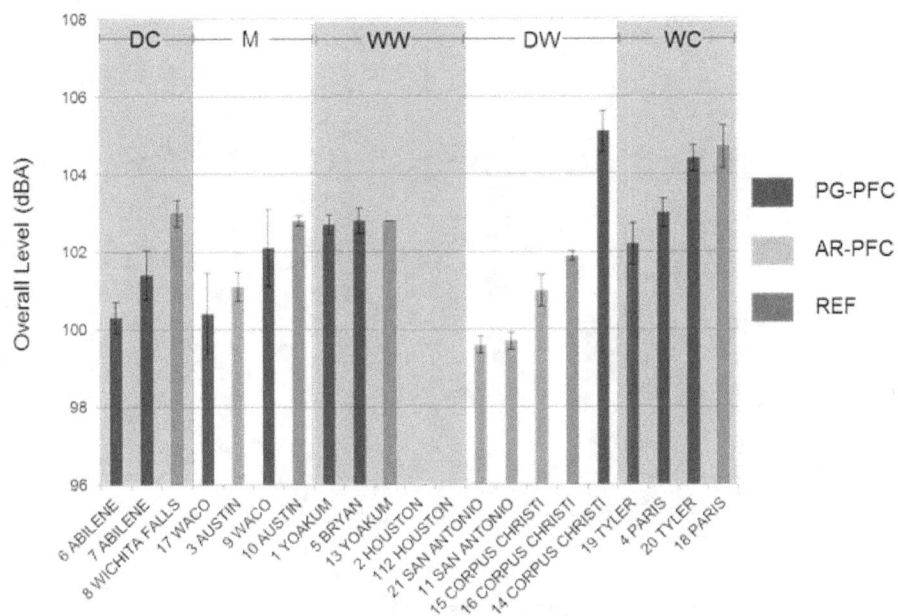

Figure 29. Effect of Climate on Noise Levels for Measurement Year 2012.

In general, AR-PFCs seem to perform acoustically well in the DW climatic region, but it may be that the quietest PFCs are AR, and happen to be located in San Antonio. There is also one fairly quiet AR-PFC in Corpus Christi, which corresponds to that climatic region as well. Reference sections appear to have higher noise levels in the cold regions. Figure 30 summarizes the OBSI noise measurements taken over the years by climatic region, showing the average throughout the life of the project. Figure 30 shows that the M region has the lowest noise levels, and that the WC region has the loudest pavements. The DC and DW regions have very similar noise levels over the years.

To further assess the differences between the noise levels obtained from the different climatic regions, the researchers performed an ANOVA with a significance level of $\alpha=0.05$ using all the measurements conducted in this project. The results, shown in Table 5, indicate that the differences in noise levels between climatic regions are statistically significant (p-value$< \alpha$); therefore, according to these numbers, the climatic classification of the pavement sections has an impact in noise measurements.

It can also be observed that the variability for measurements taken in regions classified as M and WW is very small compared to the other regions, and that the results corresponding to the DW region have a very large variability.

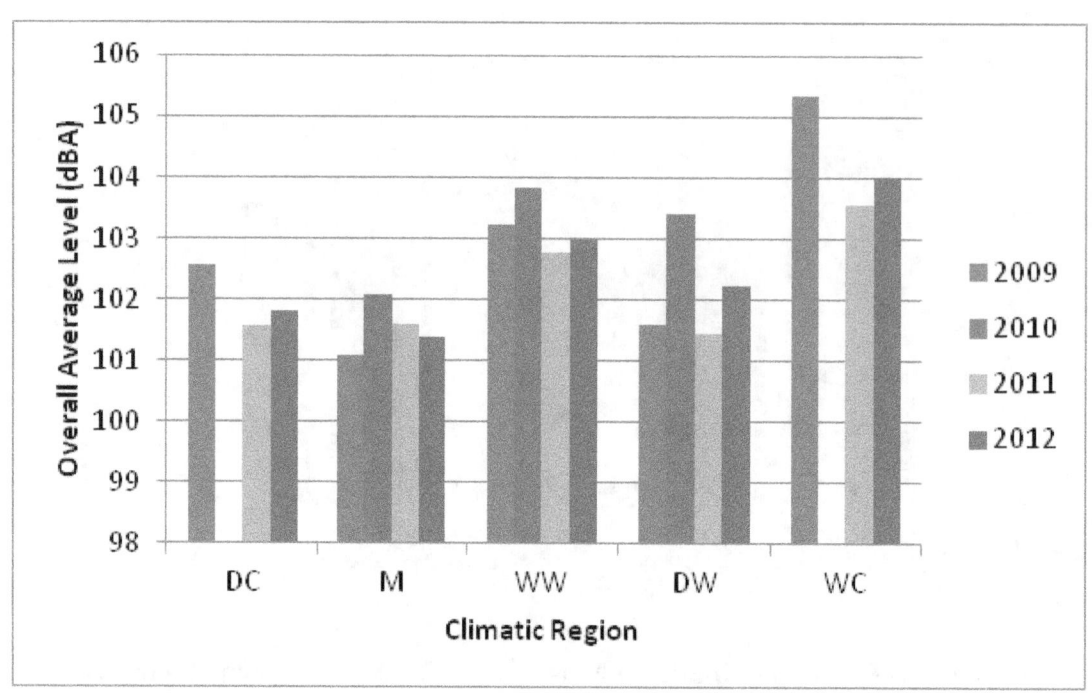

Figure 30. Average OBSI Noise Levels by Climatic Region.

Table 5. Statistical Differences in Noise Levels Based on Climatic Region.

Groups	Count	Sum	Average	Variance		
DC	9	917.79	101.98	1.442		
M	15	1523.43	101.56	0.689		
WW	14	1443.87	103.13	0.539		
DW	20	2043.44	102.17	6.019		
WC	12	1251.76	104.31	2.226		
ANOVA						
Source of Variation	SS	df	MS	F	P-value	F crit
Between Groups	62.489	4	15.6221	6.0792	0.0003	2.5130
Within Groups	167.036	65	2.5698			
Total	229.524	69				

DRAINABILITY

WFVs were acquired during the years 2009, 2010, and 2012 only on PFC pavement sections. UTBHMWC and dense-graded HMA pavement sections were not measured since they are essentially impervious. To identify the effect of traffic on drainability, researchers took OWP and BWP measurements. Results are shown in Figure 31. The data for both measurements have good correlation and align closely to the equality line (45-degree angle dashed line in Figure 31). The WFVs tend to sit slightly below the equality line, which means that the OWP WFVs are somewhat higher (as expected).

A statistical comparison between drainability measured BWP and OWP was performed using a paired two-sample t-test (two-tailed) with a significance level of 5 percent ($\alpha = 0.05$). The results show that, in general, there is no difference between the BWP and OWP WFVs taken at the same locations within the same measurement year. The exceptions are pavement sections 2, 5, and 20 in year 2009; 5 and 19 in 2010; and 5 and 15 in 2012. In addition, in the year 2009 drainability measurements were done on the shoulder in eight of the pavement sections (i.e., 1, 2, 4, 6, 9, 15, 17, and 21). In comparison to the BWP values, no statistical difference was observed between these two measurements.

In general, more variability was observed for AR-PFCs (Avg. Coefficient of Variance [COV] = 34 percent) than for PG-PFCs (Avg. COV = 18 percent). In addition, it is apparent that after a WFV of about 90 seconds, the correlation of BWP vs. OWP becomes weaker. Therefore, 90 seconds was established as the threshold to define draining PFC pavements vs. impervious ones. All OWP drainability measurements for PG-PFC pavements were below the established threshold.

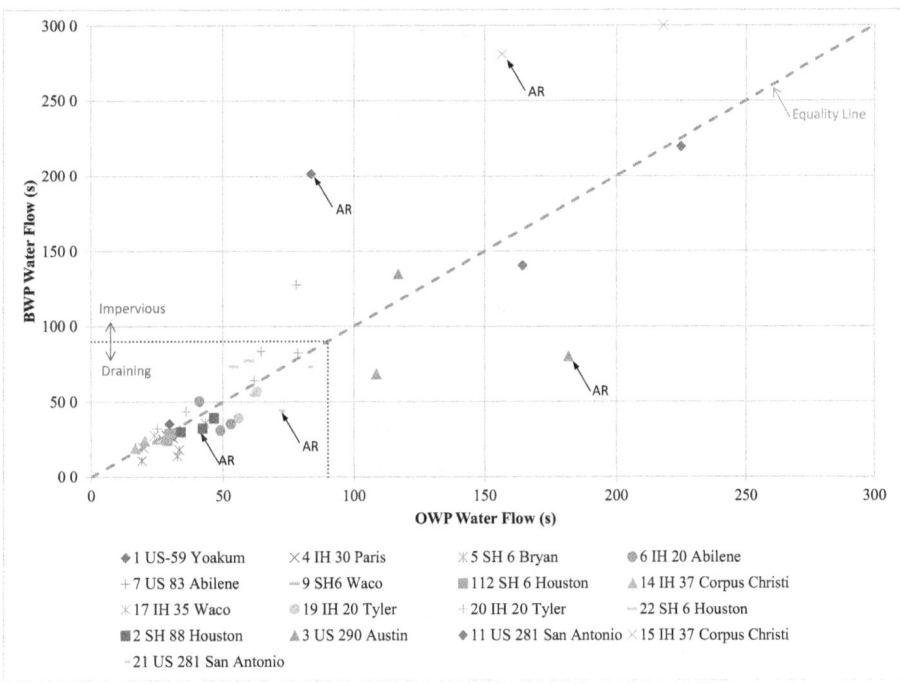

Figure 31. Effect of Traffic on Drainability.

The effect of time in drainability is illustrated in Figure 32 and Figure 33 for AR-PFCs and PG-PFCs after applying the 90-second impervious threshold criteria to the subsection measurements. Only OWP measurements are presented due to the strong correlation between

BWP and OWP values. In Figure 32, the values are shown based on the duration of the project, while in Figure 33, the results are presented taking into account the age of the pavement sections.

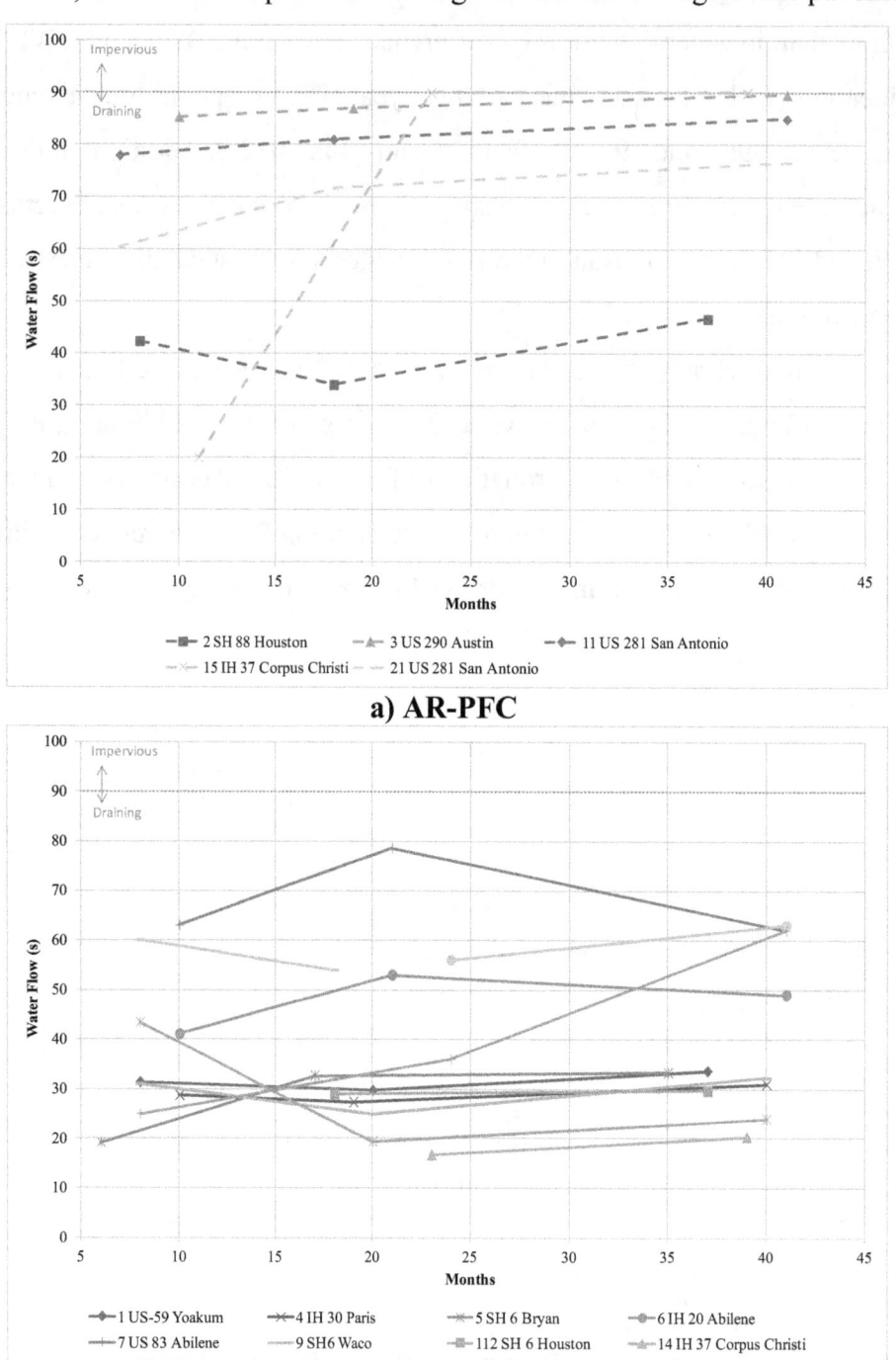

a) AR-PFC

b) PG-PFC

Figure 32. OWP Drainability Measurements throughout the Life of the Project.

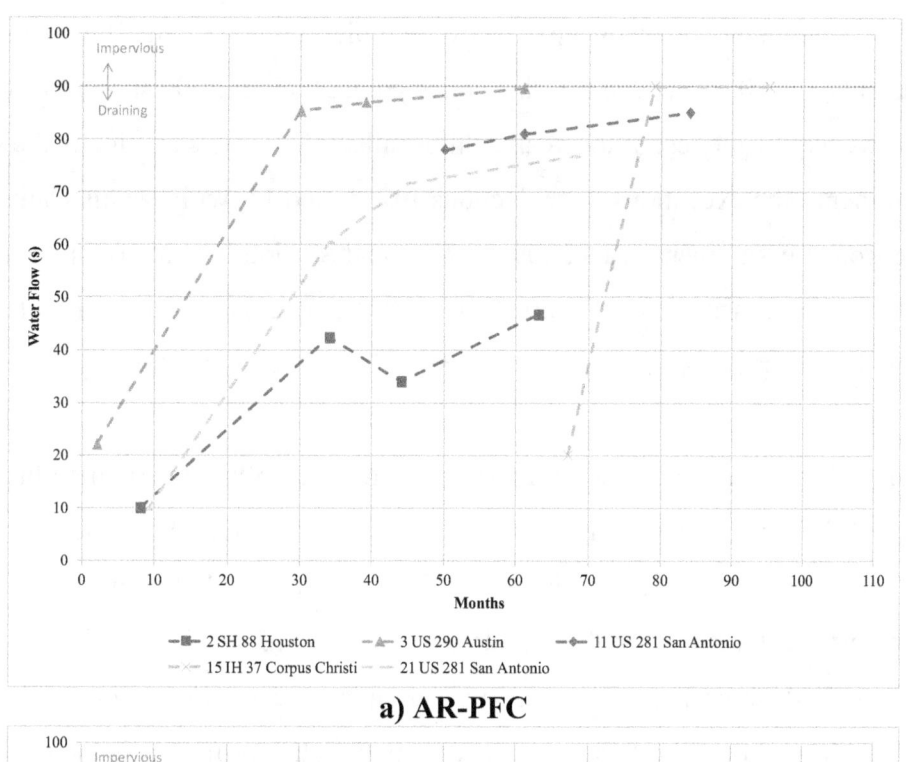

a) AR-PFC

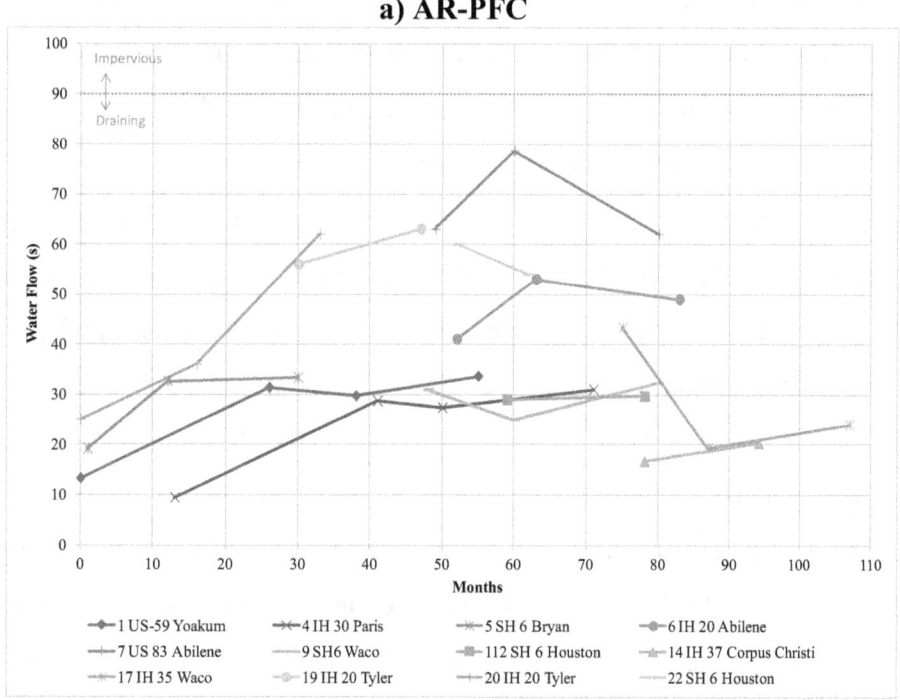

b) PG-PFC

Figure 33. OWP Drainability Measurements with Pavement Life.

In Figure 33, WFVs from five field sections (i.e., 1, 2, 3, 4, and 21) that were collected in 2007 soon after construction as part of previous TxDOT Project 0-5262 *Optimizing the Design of Permeable Friction Courses (PFC)* are also included. All 2007 values were below 20 seconds.

A statistical comparison using a paired two-sample t-test (two-tailed) with a significance level of 5 percent ($\alpha = 0.05$) between the data collected in 2007 vs. 2009 showed a significant difference. However, for the rest of the years, almost all WFVs were statistically the same. For OWP measurements, the exceptions were Sections 1, 2, 5, and 15, with a statistically significant WFV change from measurement years 2009 vs. 2010 and Section 20 with a statistically significant change from measurement years 2010 vs. 2012. This shows that WFVs tend to increase when the pavement is new and then remain relatively constant throughout the life of the pavement.

For the PG-PFC pavements, after the initial increase in WFV, the drainability of more than one-third of the pavement sections remained constant at around 30 seconds (see Figure 33b). On the other hand, the WFVs for three out of the five AR-PFC sections were close to or beyond the impervious threshold of 90 seconds. Section 15, IH 37 in Corpus Christi, had a fog seal treatment applied between the 1st and 2nd measurement years, which explains the steep increase in WFV. Based on the measurements performed in this project, it is likely that the AV structure of the AR-PFC pavements is more prone to clogging, possibly due to the finer gradation and higher binder content that TxDOT specifications required.

With regard to type of binder used in the PFC pavements, the statistical comparison between the AR-PFC and PG-PFC WFVs revealed differences. On average, considering only the 5-minute field threshold on the subsection measurements, the WFV for the AR-PFC pavements increased almost twofold from 63 seconds in 2009 to 122 seconds in 2010, and then stabilized at around 138 seconds in 2012. By contrast, the average WFV for PG-PFCs remained constant at 38 seconds in 2009 and 2010, and 40 seconds in 2012 (see Figure 34). When the 90-second impervious threshold criteria was applied to the subsection measurements, the average WFVs for AR-PFCs changed from 57 to 73 and to 78 seconds, while for PG-PFCs the average WFVs were 36, 38, and 40 seconds for years 2009, 2010, and 2010, respectively (see Figure 35).

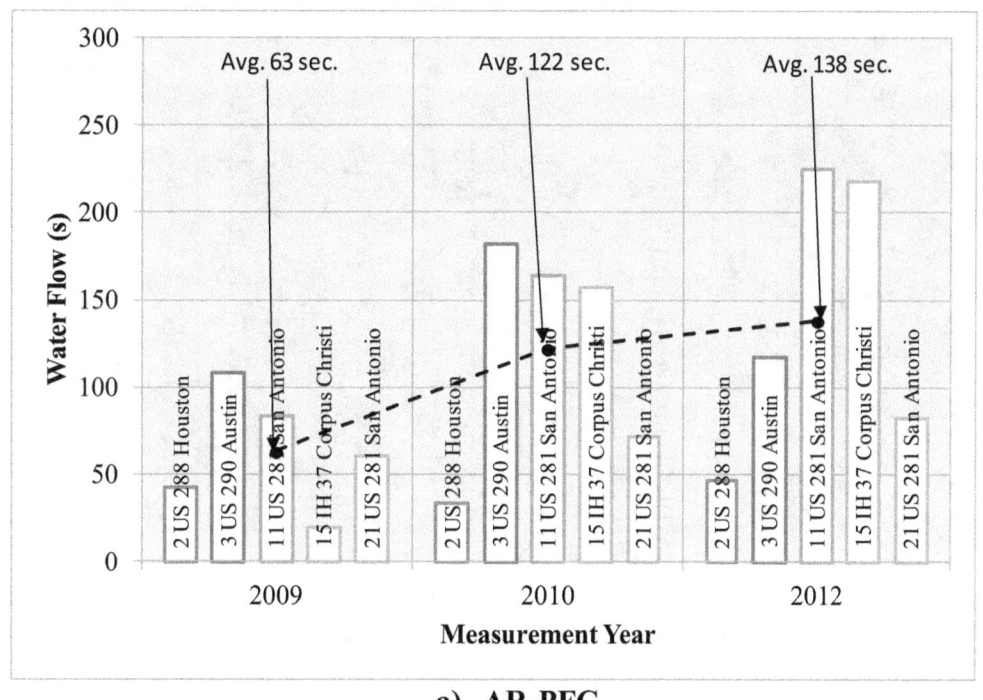

a) AR-PFC

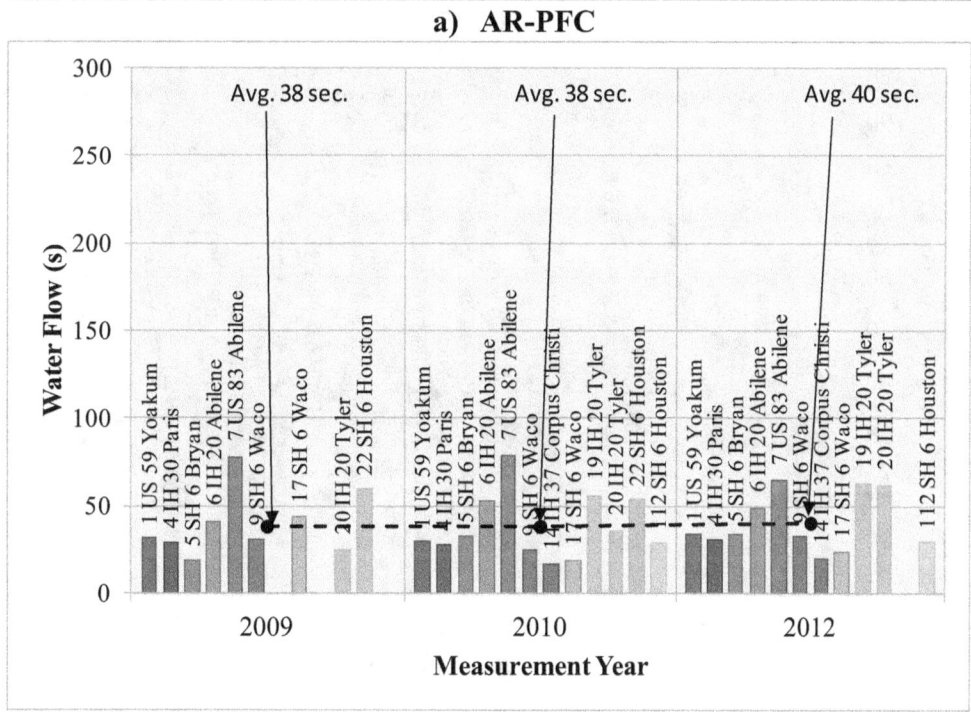

b) PG-PFC

Figure 34. OWP Drainability for Each Measurement Year with 5-Minute Impervious Field Threshold.

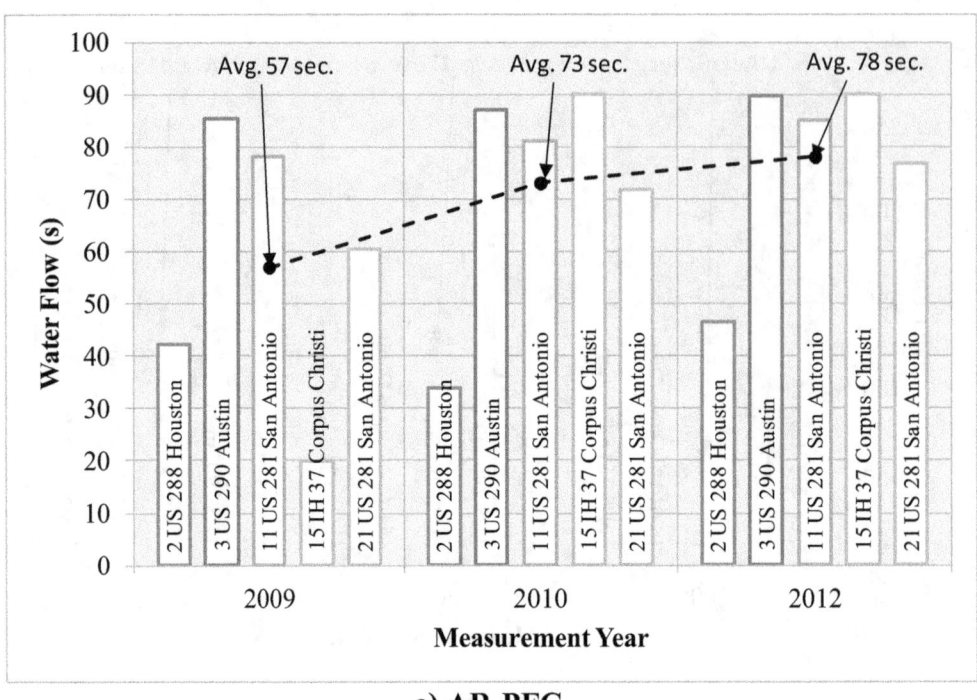

a) AR-PFC

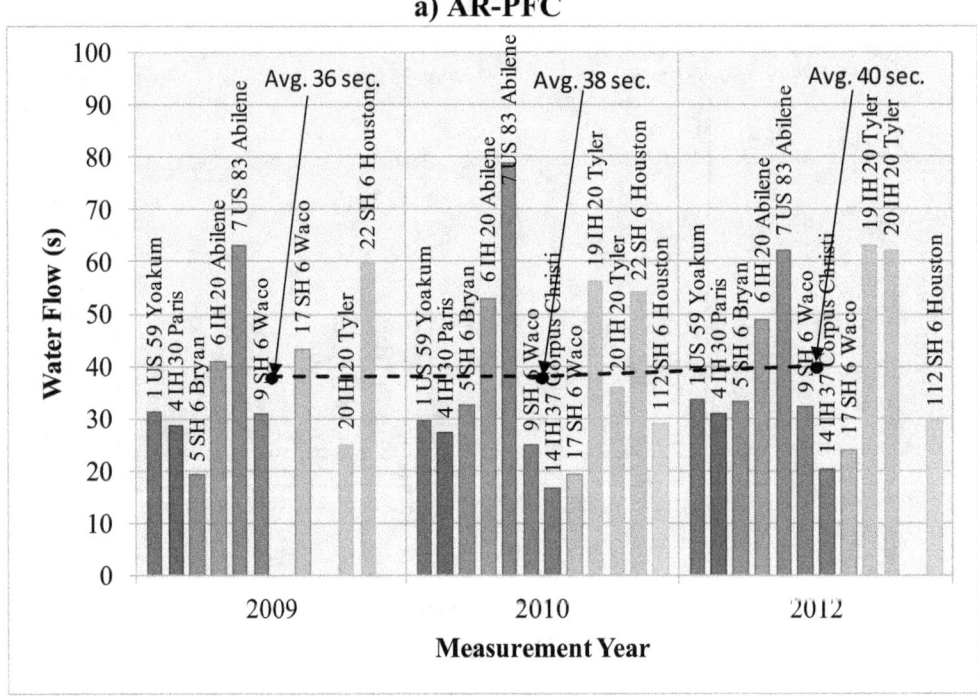

b) PG-PFC

Figure 35. OWP Drainability for Each Measurement Year with 90-Second Impervious Threshold.

The effect of aggregate classification on drainability was also assessed. Some mixtures employed aggregates classified per TxDOT's Surface Aggregate Classification (SAC) system as SAC-A or SAC-B. More importantly, a few pavement sections used a combination of SAC-A and SAC-B aggregates. Aggregates classified as SAC-A consist of high-quality materials, resistant to polishing, and having higher soundness value as compared to aggregates classified as SAC-B. Some districts have demonstrated concerns about combining SAC-A and SAC-B aggregates because they have observed crushing of the SAC-B aggregate under loading. For PFCs, this could translate into a collapse of the AV structure and subsequent loss in functionality (i.e., permeability and noise reduction).

A statistical comparison of the WFVs obtained for the pavement sections with SAC-A, SAC-B, and combination of SAC-A/B aggregates showed no significant differences. Figure 36 illustrates how the WFVs for the different type of aggregates are not clustered. The averages for each measurement year and for all years combined are also shown in Figure 36.

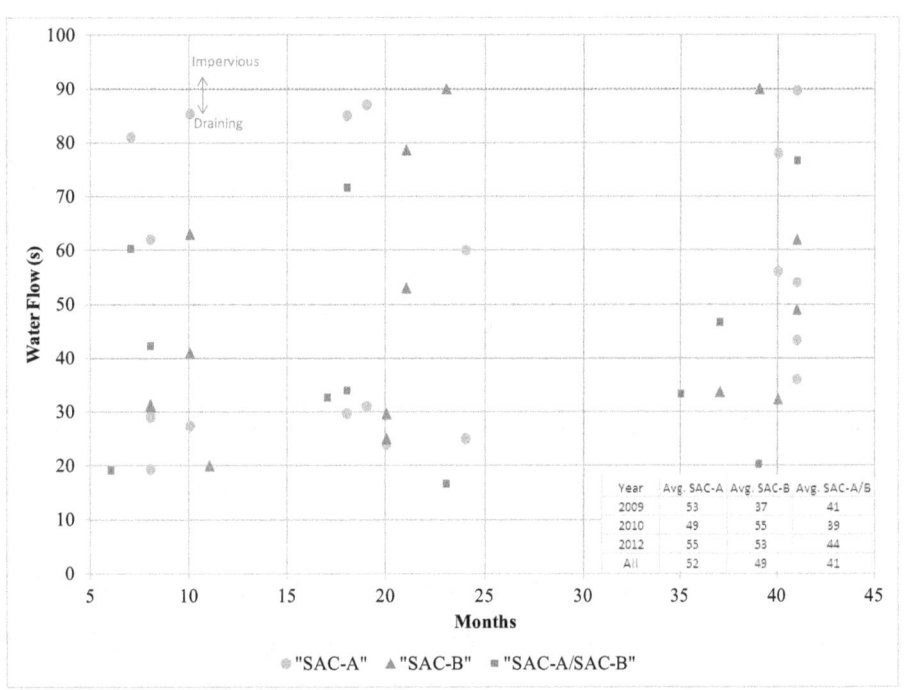

Figure 36. Effect of Aggregate Classification on Drainability OWP Measurements.

Another factor that was explored was the effect of climate on drainability. Figure 37 shows the results for each measurement year. A statistical analysis using ANOVA and Tukey's HSD resulted in no statistical significant differences for measurement years 2009 and 2010. However, in year 2012 as well as when all three years were analyzed together, the WW and DW climatic regions were statistically different from each other. The WFVs in the WW climate were significantly lower when compared to the other climatic regions, around 35 seconds in average, while the WFVs in the DW climate were much larger, around 63 seconds. The WFVs for the other three climatic regions arranged in increasing order as follows: WC with an average of about 40 seconds, M with an average of almost 50 seconds, and DC with an average of 58 seconds. This seems to indicate that the amount of rainfall in a particular climatic region does play an important role in the continued drainability of PFCs, especially in warm climates.

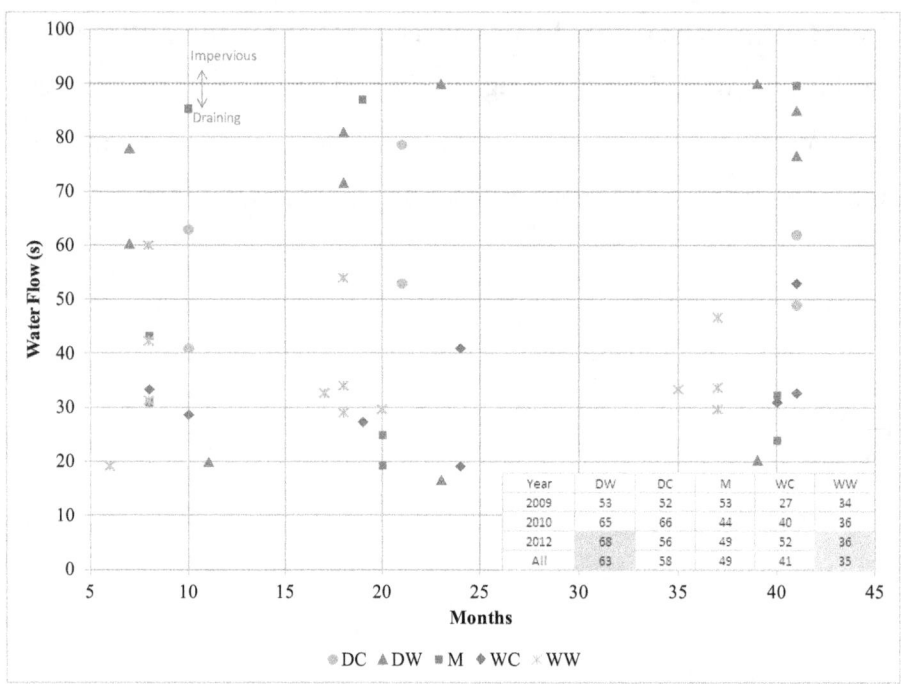

Figure 37. Effect of Climate on Drainability OWP Measurements.

TEXTURE

Texture measurements were performed on all pavement sections in years 2009, 2010, and 2012. The effect of traffic on texture was assessed by comparing OWP vs. BWP MPD values. Figure 38 illustrates the relationship between these measurements. There is a strong correlation between the MPD values acquired OWP and BWP for both PFC and non-PFC pavements. For PFCs, the MPD values acquired BWP align slightly above the equality line, which implies that texture BWP is somewhat larger than texture OWP (see Figure 38a). The average MPD BWP is 1.59 mm (0.063 inch) and 1.47 mm (0.057 inch) for OWP.

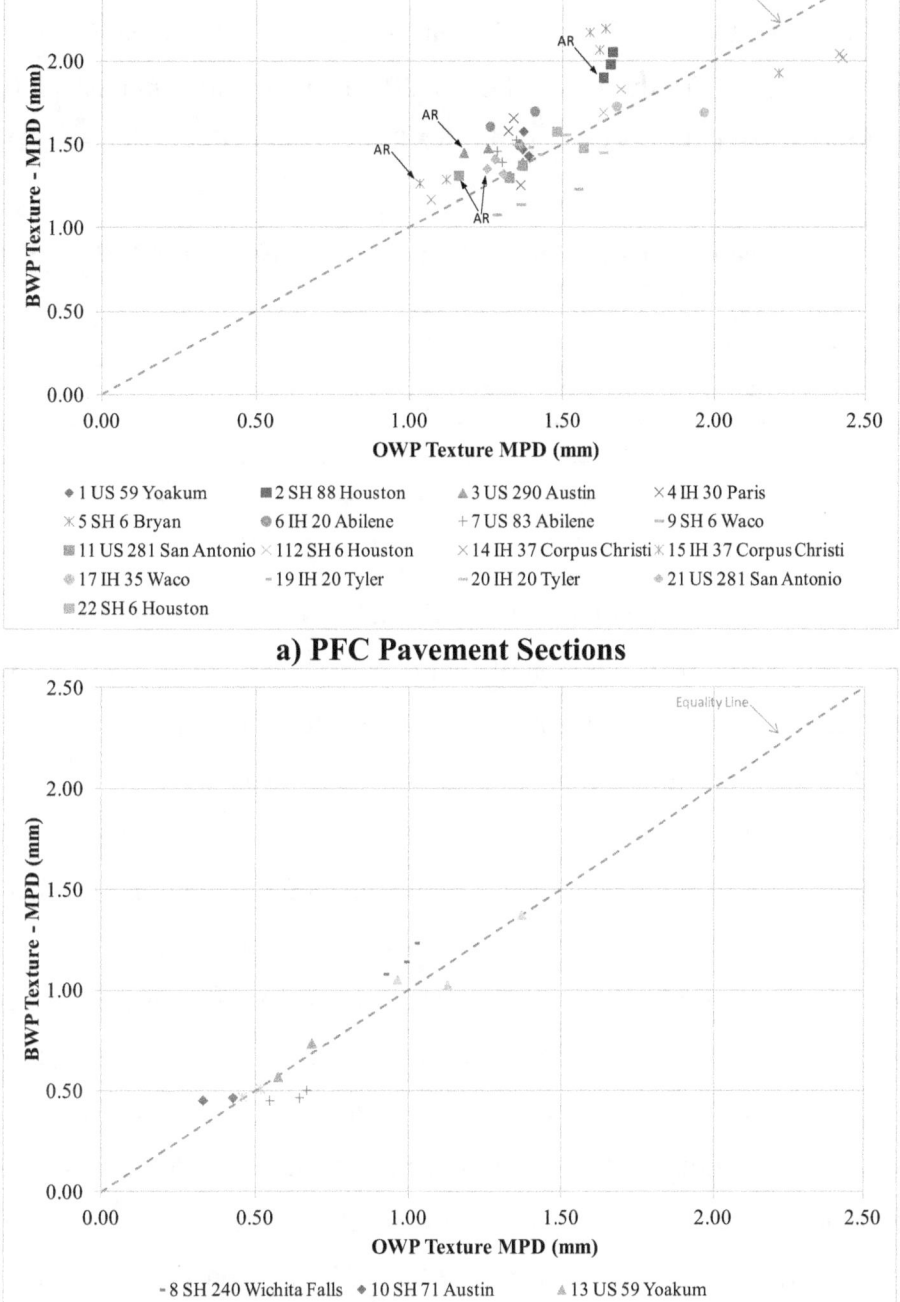

Figure 38. Effect of Traffic on Texture Measurements.

A statistical comparison between texture measured BWP and OWP was performed using a paired two-sample t-test (two-tailed) with a significance level of 5 percent ($\alpha = 0.05$). Sections with statistical significant different MPD values taken at the same locations within the same measurement year are listed in Table 6. Differences between texture measurements acquired OWP vs. BWP were greater for PFCs; the other pavement types showed only a few sections with statistical differences in MPD values. For PFCs, the BWP texture measurements for PFCs were more variable (Avg. COV 6.5 percent) than the OWP MPD (Avg. COV 5.2 percent).

Table 6. Pavement Sections with Statistical Significant Differences MPD OWP vs. BWP.

Pavement Type	Measurement Year		
	2009	2010	2012
AR-PFC	2, 15, 21	2, 3	15
PG-PFC	5, 6	14	5, 6, 17
TBPFC	4	4	
Dense HMA	10		18
UTBHMWC	8		

The change of texture with time is illustrated in Figure 39 and Figure 40. Figure 39 shows the data with respect to the duration of the project, while Figure 40 takes into account the age of the pavement. All pavement sections with the exception of section 14 and 15 in Corpus Christi show relatively flat texture trends. A statistical comparison using a paired two-sample t-test (two-tailed) with a significance level of 5 percent ($\alpha = 0.05$) for OWP texture values showed significant differences between MPD values collected in 2009 vs. 2010 for only these two sections. The rest of the sections resulted in no statistical significant differences.

In the case of section 15, a fog seal was applied between the measurements performed in years 2009 and 2010, causing the significant decrease in texture. For Section 14, the raveling experienced by the pavement towards the end of its service life more likely has caused the increase in texture. In addition, this pavement section was laid on top of an existing surface with considerable amounts of cracking. Also, between years 2009 and 2010, several freeze/thaw cycles were experienced in that region. Sections 14 and 15 were considered extreme cases and excluded from the subsequent analysis.

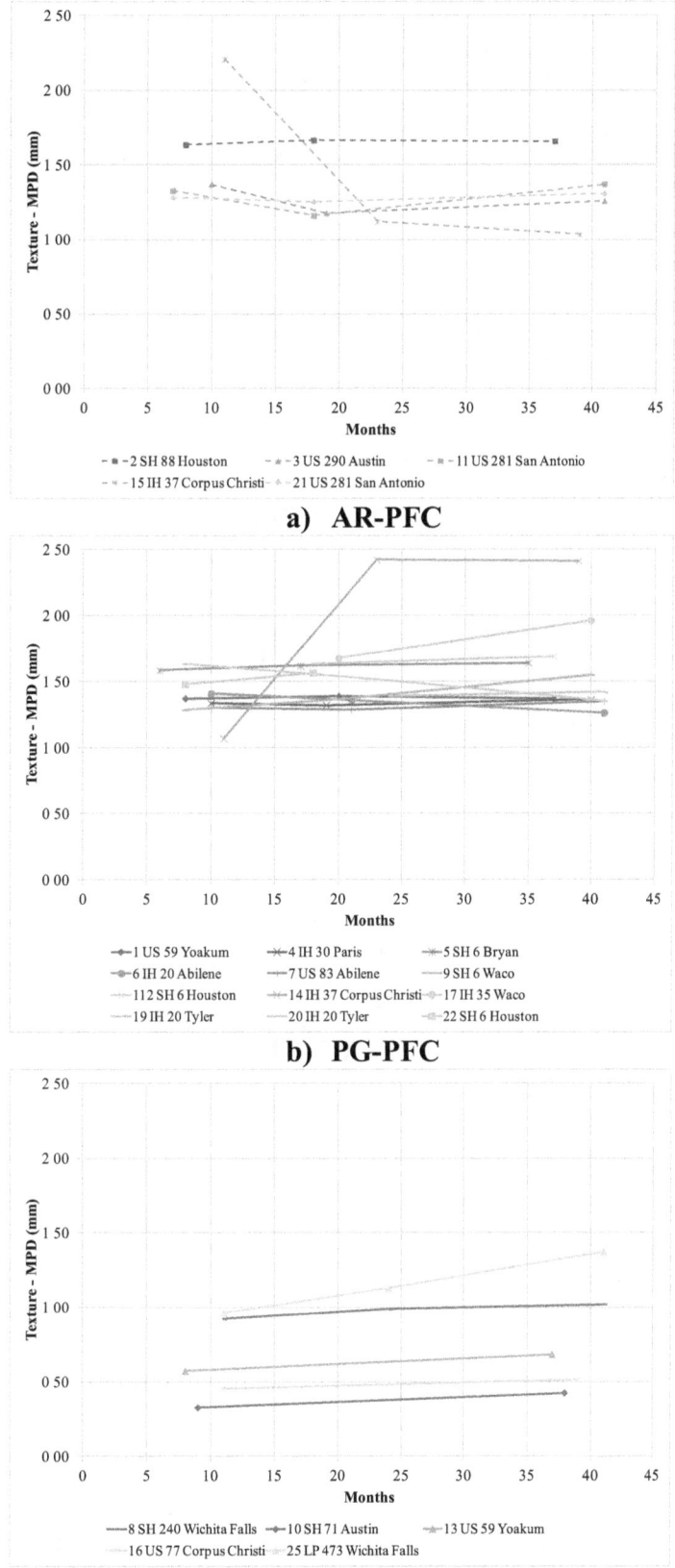

a) AR-PFC

b) PG-PFC

c) Non-PFC Pavement Sections

Figure 39. OWP Texture Measurements throughout the Life of the Project.

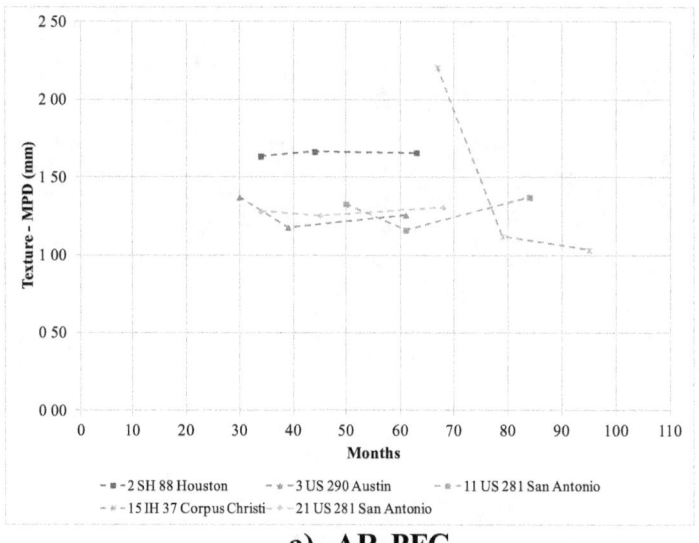

a) AR-PFC

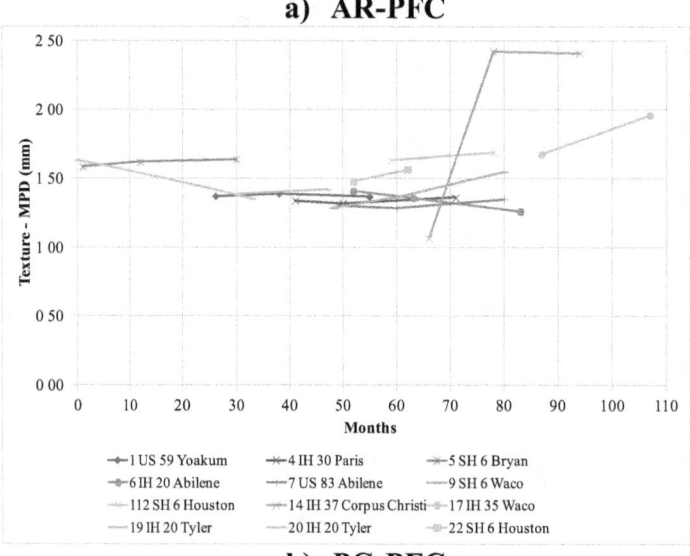

b) PG-PFC

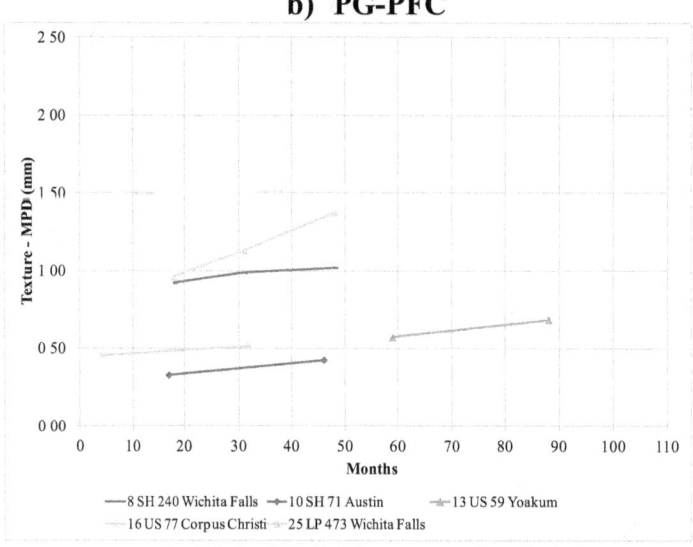

c) Non-PFC Pavement Sections

Figure 40. OWP Texture Measurements with Pavement Life.

Figure 41 illustrates the effect of binder/mixture type. A statistical comparison of the results using ANOVA and Tukey's HSD showed that both AR and PG-PFCs had a statistical equivalent texture and superior than the dense-graded HMA pavement sections used as reference. The UTBHMWC texture values aligned between the PFCs and HMAs. The statistical results for the mixtures and average texture values are presented in table format in Figure 41 with color-coding indicating the mixtures that belong to the same statistical group. Texture values for AR-PFCs were statistically equivalent to both PG-PFCs and UTBHMWC in 2012. These results indicate that PFCs offer better texture characteristics than conventional dense-graded HMA and the texture values stay somewhat constant over time.

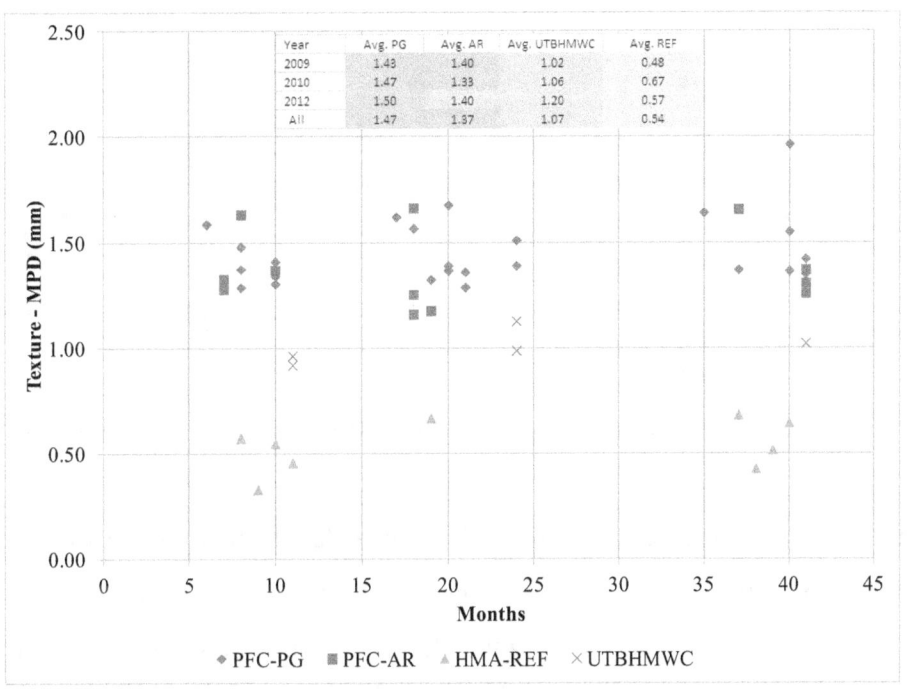

Figure 41. Effect of Binder/Mixture Type on OWP Texture Measurements

The effect of aggregate classification on PFC texture was also evaluated. Results show that the average texture for the PFC sections with SAC-A aggregates were around 1.45 mm; (0.057 inch), for SAC-B aggregates 1.36 mm (0.053 inch); and for a combination of SAC-A/B aggregates, 1.52 mm (0.060 inch) (see Figure 42). For each measurement year, the difference in MPD values was statistically insignificant. When results for all years were combined, the texture values for the sections with SAC-A/B aggregates were statistically different from the SAC-B aggregates, while the texture values for the sections with SAC-A aggregates was statistically equivalent to sections with either SAC-B or SAC-A/B aggregates. Overall, one can assume that the use of aggregates with different classifications in the PFC mixtures has no significant impact on the texture values. Moreover, texture on PFC pavement sections seems to remain unchanged over time.

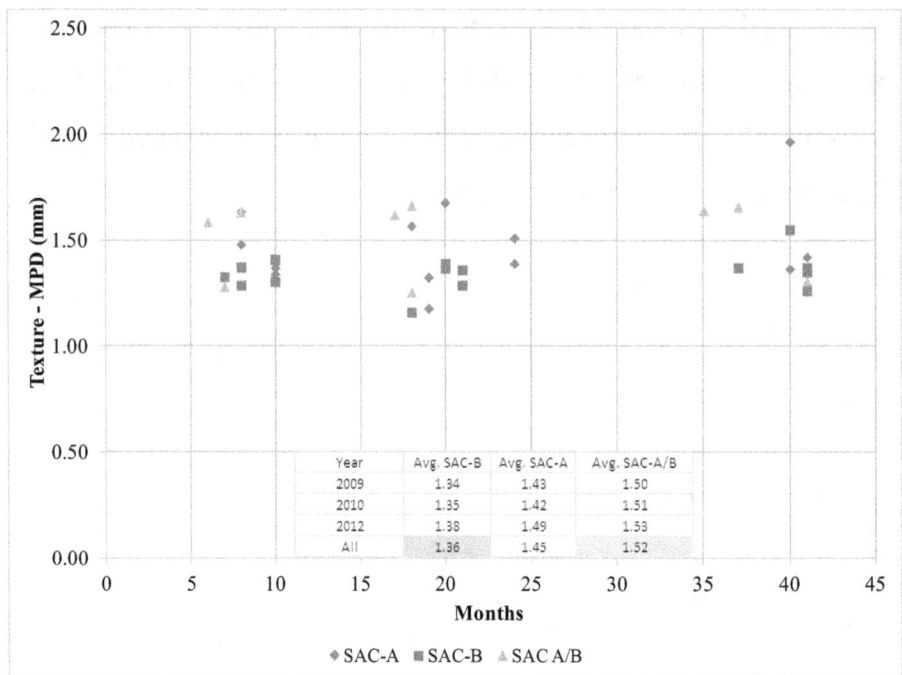

Figure 42. Effect of Aggregate Classification on OWP Texture Measurements for PFC Pavements.

With respect to the effect of climate on texture, a statistical analysis using ANOVA and Tukey's HSD was performed, taking into account the climatic region of the PFC pavement sections. The climatic regions with significant differences are color-shaded in the table included in Figure 43. The largest texture values corresponded to the pavement sections located in the WW climatic region, while the lowest texture was measured in the sections located in the DC and DW climatic regions. These differences are consistent with each measurement year, and as previously discussed, texture values remain unchanged with time. This indicates that the texture achieved in PFCs soon after construction is more critical than the climatic region where the pavement section is located.

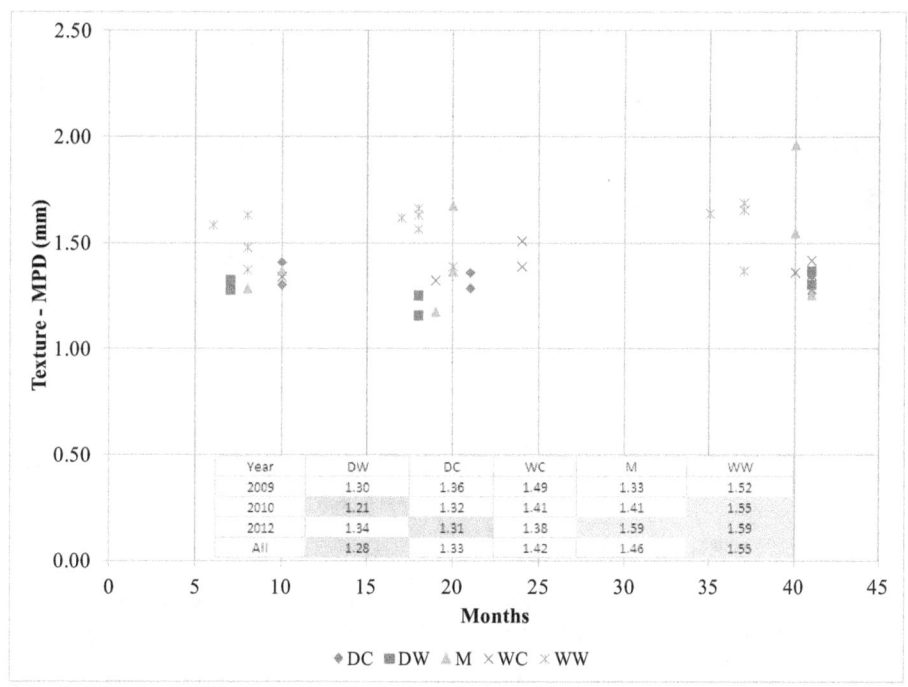

Figure 43. Effect of Climate on Texture OWP Measurements for PFC Pavements.

FRICTION

An important characteristic of PFCs is their ability to provide adequate friction in order to aid in reducing the number of accidents, especially during wet weather events. The effect of traffic on friction was evaluated using OWP and BWP DFT measurements acquired at 60kph. Figure 44 illustrates the comparison. There was a strong correlation between the OWP and BWP, which aligned close to the 45-degree equality line. The friction values BWP were slightly higher BWP than OWP, especially for non-PFC pavements (Figure 44b). For PFCs, the average friction at 60 kph BWP was 0.42 and 0.39 for OWP, while for non-PFCs the average friction BWP and OWP was 0.44 and 0.39, respectively.

A statistical comparison between texture measured BWP and OWP was performed using a paired two-sample t-test (two-tailed) with a significance level of 5 percent ($\alpha = 0.05$). Sections with statistical significant different friction values taken at the same locations within the same measurement year are listed in Table 7. Very few sections resulted in statistical significant differences, and the majority of these occurred in the first measurement (i.e., 2009). For PFCs, the BWP friction measurements for PFCs were slightly more variable (Avg. COV 10.0 percent) than OWP (Avg. COV 9.1 percent).

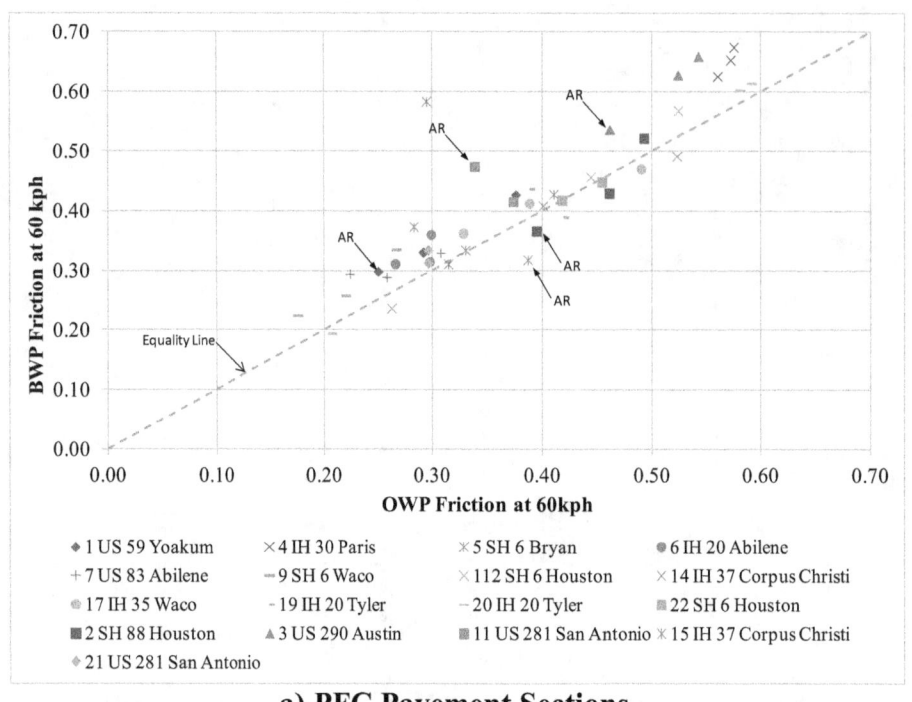

a) PFC Pavement Sections

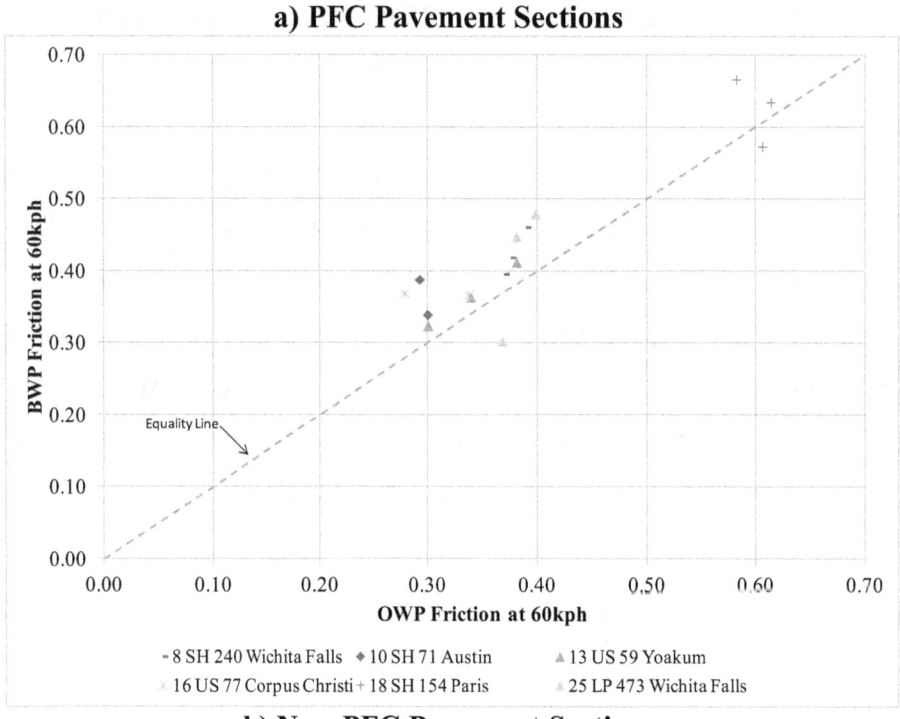

b) Non-PFC Pavement Sections

Figure 44. Effect of Traffic on Friction Measurements.

Table 7. Pavement Sections with Significant Differences in Texture OWP vs. BWP.

Pavement Type	Measurement Year		
	2009	2010	2012
AR-PFC	3	3	11
PG-PFC	9, 14	9	1, 14
TBPFC	4		
Dense HMA	10, 16		
UTBHMWC	8		

The change in friction with time is illustrated in Figure 45 and Figure 46. Figure 45 illustrates the trends with respect to the duration of the project, while the friction measurements in Figure 46 are plotted against the life of the pavement sections. In general, PG-PFCs show upward friction trends, while AR-PFCs and non-PFC pavement sections show flat or downward friction trends with time. A statistical comparison using a paired two-sample t-test (two-tailed) with a significance level of 5 percent ($\alpha = 0.05$) was done to compare the friction data acquired during the different measurement years. For PFCs, sections with statistically significant OWP friction values from measurement years 2009 vs. 2010 included sections 5, 6, 14, and 20, and from measurement years 2010 vs. 2012 sections 7 and 14. For non-PFC pavements, only section 13 showed statistical differences.

As previously mentioned, Section 14 experienced raveling, which could explain the significant increase in measured texture. Section 15, which was treated with a fog seal between years 2009 and 2010, showed a decrease in friction, although not statistically significant. The significant increase in friction observed in pavement Section 20 is likely due to the fact that this pavement section was constructed just before the first measurement was acquired, and it is known that after traffic wears down the asphalt binder at the surface of the pavement the friction values tend to increase.

The effect of binder/mixture type in friction was also quantified. Figure 47 shows the comparison. ANOVA and Tukey's HSD analysis showed that the texture values for all mixtures were statistically equivalent. Based on the average friction values included in the summary table in Figure 47, it is apparent that the friction measurements had more variability year to year. Overall the values did not cluster in distinct groups and stayed somewhat constant with time.

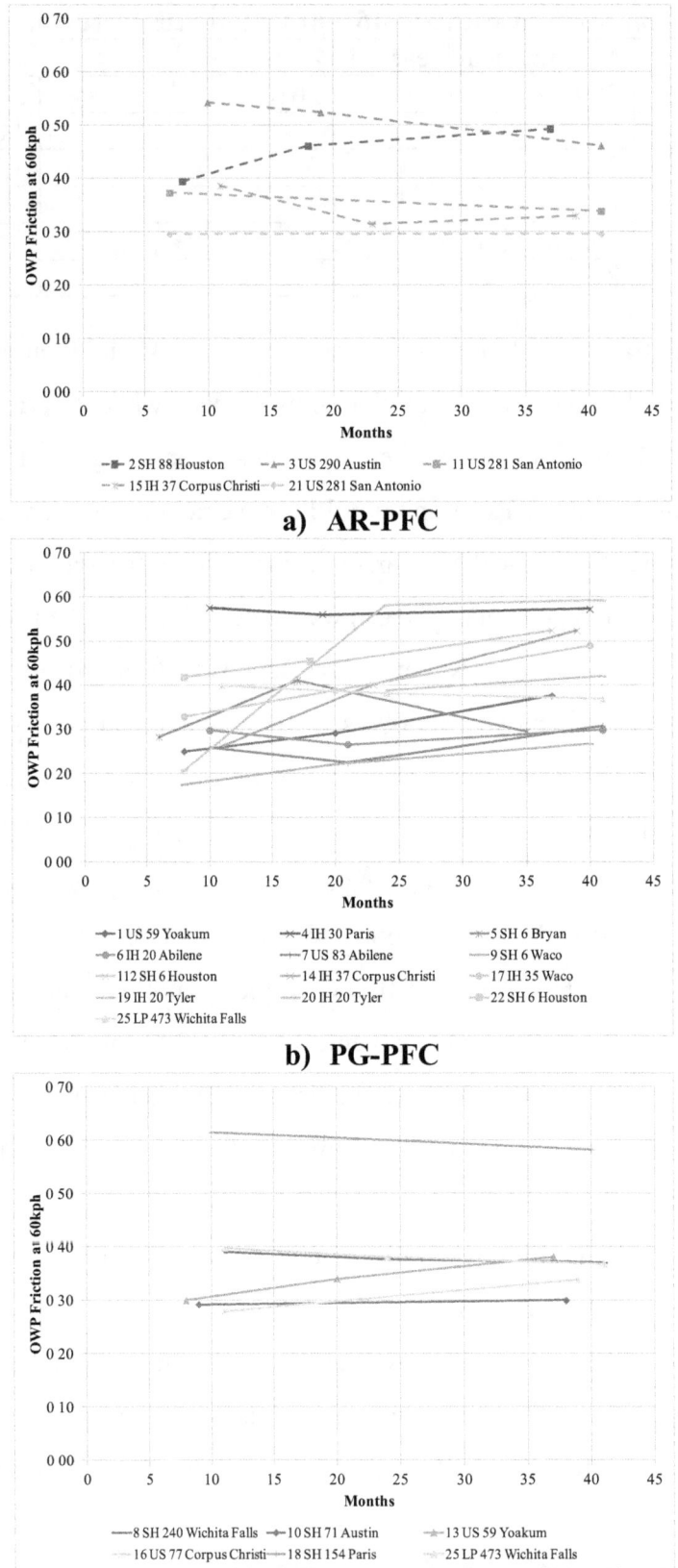

a) AR-PFC

b) PG-PFC

c) Non-PFC Pavement Sections

Figure 45. OWP Friction Measurements throughout the Life of the Project.

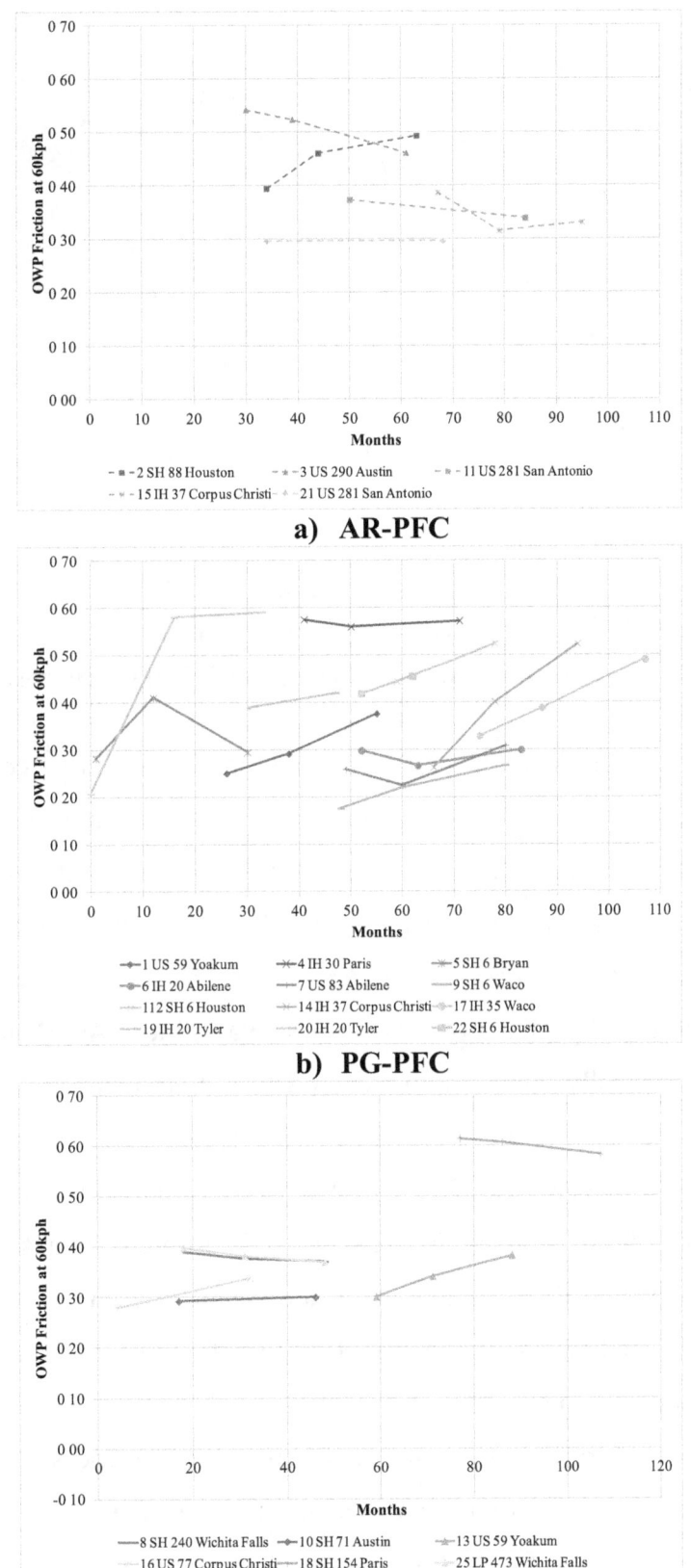

a) AR-PFC

b) PG-PFC

c) Non-PFC Pavement Sections

Figure 46. OWP Friction Measurements with Pavement Life.

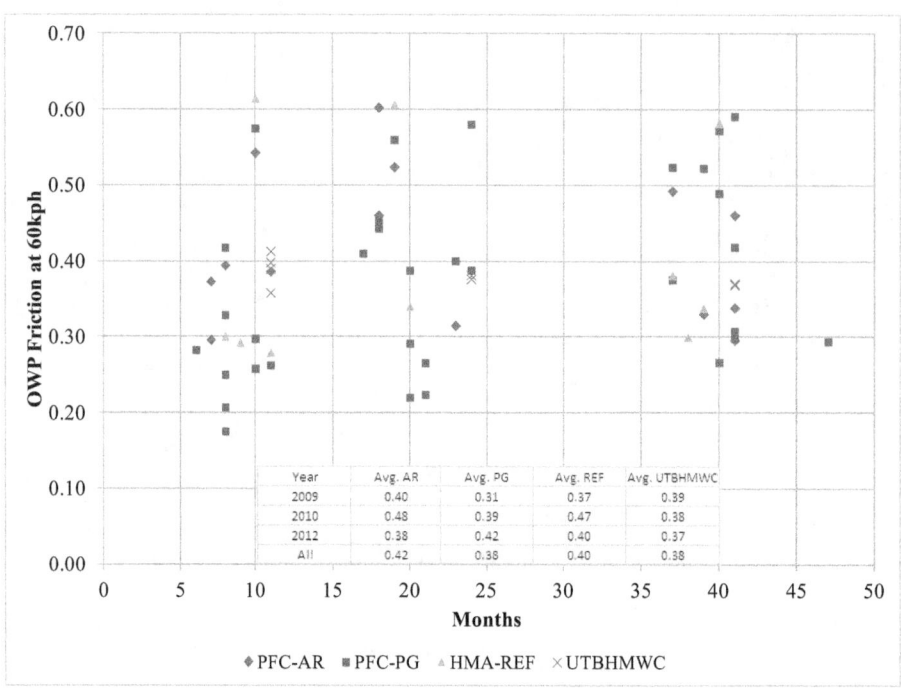

Figure 47. Effect of Binder/Mixture Type on OWP Friction Measurements.

The effect of aggregate classification on texture for PFCs is illustrated in Figure 48. Results with statistical significant differences are indicated with different color shading in the summary table included in Figure 48. As can be observed, pavement sections with SAC-B aggregates had statistically significant lower friction values as compared to those pavement sections employing either SAC-A or SAC-A/B aggregates. The friction values for the sections with SAC-A/B aggregates align in between the values obtained for SAC-B and SAC-B aggregates and are statistically equivalent to section with SAC-B aggregates in 2009 and 2012. In addition, sections with SAC-A/B aggregates had friction values equivalent to those sections employing SAC-A aggregates in year 2012. Pavement sections employing exclusively SAC-A or SAC-B aggregates have friction values that are statistically different all measurement years.

To discard the possibility of lower friction values being caused by larger volumes of traffic, researchers obtained the Average Annual Daily Traffic (AADT) from PMIS for the available pavement sections. Figure 49 shows the AADT values. Sections with SAC-B aggregates have the lowest traffic levels. This verifies that the differences in friction are indeed a function of the aggregate classification and not due to variations in traffic level.

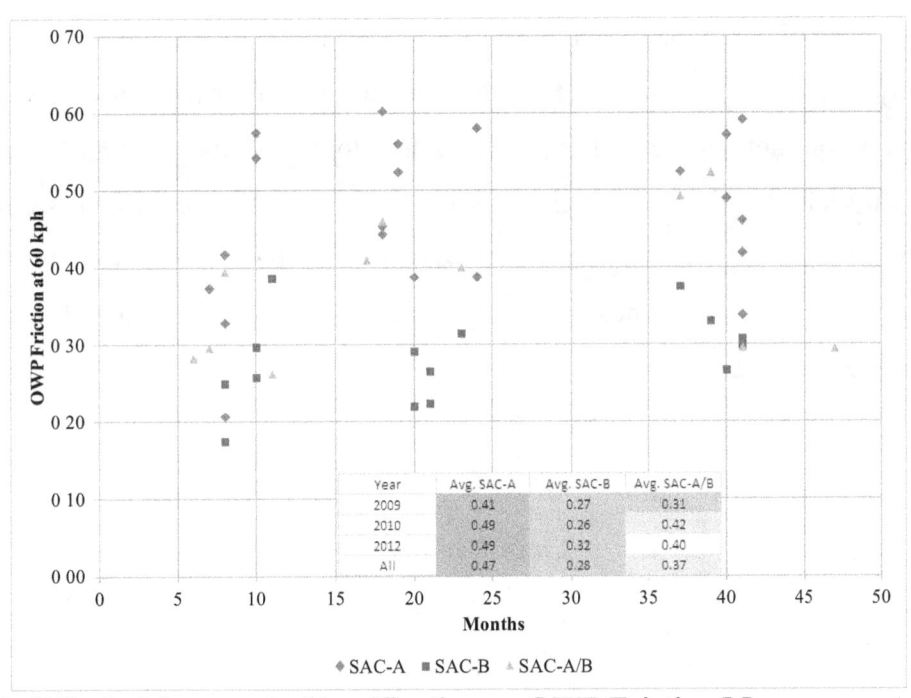

Figure 48. Effect of Aggregate Classification on OWP Friction Measurements for PFC Pavements.

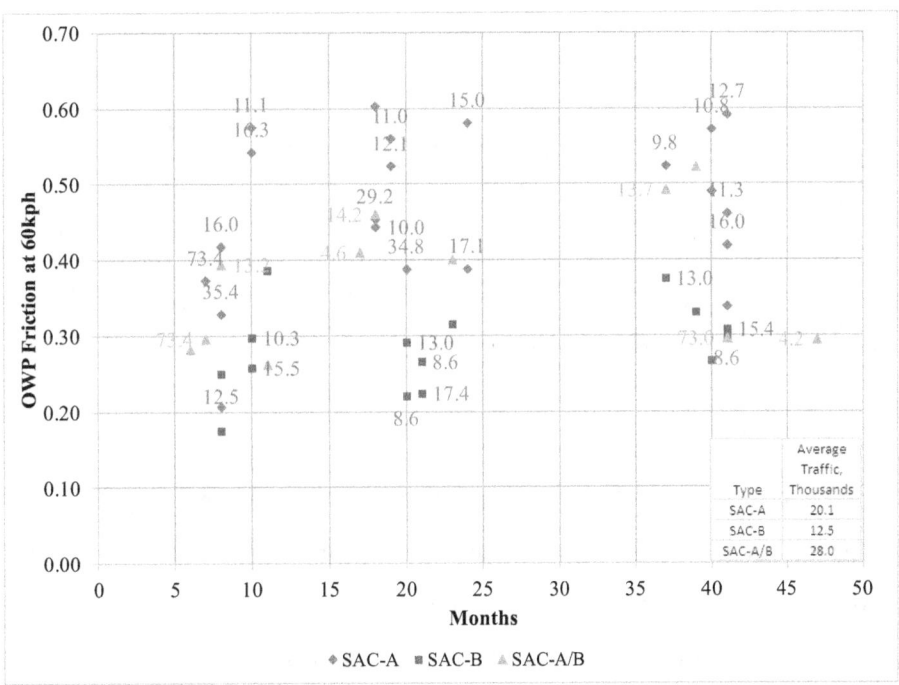

Figure 49. Effect of AADT on Aggregate Classification and Their Effect on OWP Friction Measurements for PFC Pavements.

The last factor analyzed was climate using ANOVA and Tukey's HSD. The results of the statistical analysis are summarized in the table included in Figure 50. Results that are statistically different from each other are shaded with different colors. The cells that are not shaded are statistically equivalent to other climates. The dry climates had lower friction values, which stayed relatively constant with time. The friction values for the M climatic region were in between the values for the dry and wet climates and had a small but steady increase with time. The WC climatic region had the highest friction values. Both the WC and WW climatic regions have an increase in friction from measurement years 2009 vs. 2010; subsequently, the values remained relatively constant.

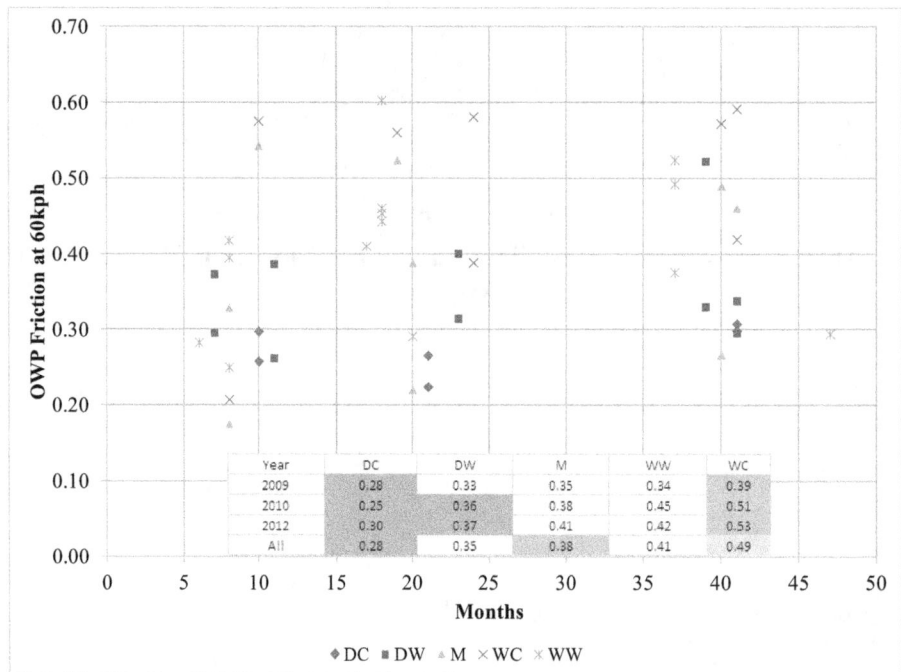

Figure 50. Effect of Climate on Friction OWP Measurements for PFC Pavements.

SKID

TxDOT extracted the skid numbers for the pavement sections from the PMIS database. Not all values were available for all sections and/or all years; Figure 51 presents the available data. As compared to the WFV, MPD, and friction measurements, the SNs seem to have more variability and not follow a consistent trend. For some sections, the values increase; for others, decrease; and some others stay constant.

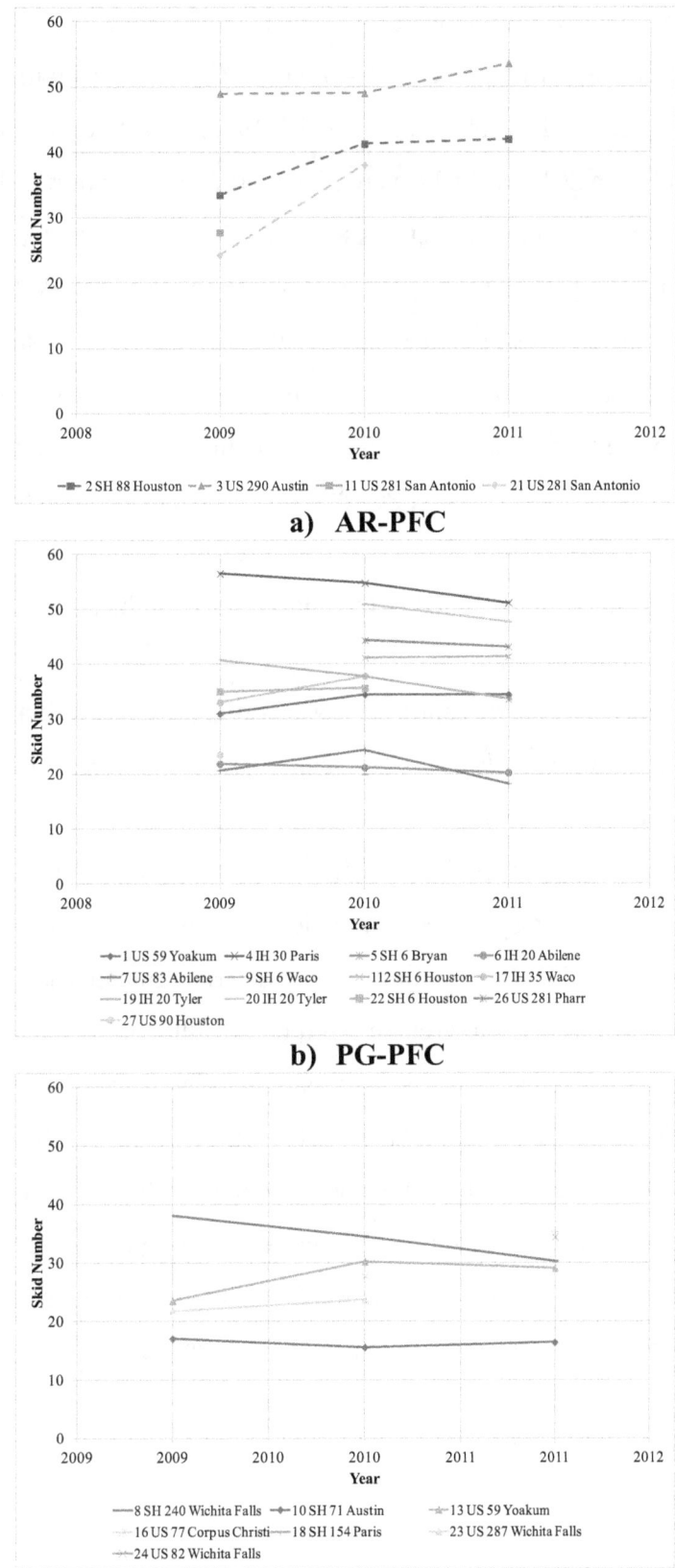

a) AR-PFC

b) PG-PFC

c) Non-PFC Pavement Sections
Figure 51. SN Measurements with Time.

The differences in SN with time were assessed using a paired two-sample t-test (two-tailed) with a significance level of 5 percent ($\alpha = 0.05$). PFC pavements with significant changes between 2009 vs. 2010 included sections 1, 2, 3, 7, 17, 19, and 21. Sections with statistical significant changes between 2010 vs. 2011 were 3, 4, 7, 19, and 20. For non-PFC sections, significant changes were noted only in Sections 8 and 13.

The effect of binder/mixture type in SN was also evaluated. ANOVA and Tukey's HSD were used to assess the differences in values. The results are shown in the summary table included in Figure 52. Sections with statistically significant differences are shaded with distinct color; the cells with no shading were statistically equivalent to one or more mixture types. From the results, it is apparent that AR-PFCs and PG-PFCs have higher SN values compared to both UTBHMWC and dense-graded HMA, with AR-PFCs having the largest SN values. When only PFCs and dense-graded HMA are included in the analysis, PFCs cluster together and are statistically different from HMA.

The effect of aggregate classification on SN was quantified as well as illustrated in Figure 53. ANOVA and Tukey's HSD resulted in statistical significant differences between all aggregate classes (except in 2011 where SAC-A and SAC-A/B were statistically equivalent). Sections employing SAC-A aggregates had the highest SNs, while sections in which mixtures were comprised of only SAC-B aggregates had the lowest SNs. This confirms the friction observations, that also showed that the sections with SAC-B aggregates had the lowest friction as opposed to the sections with SAC-A aggregates, and that resulted in the largest friction values.

Finally, the effect of climate was assessed using the SN measurements. Figure 54 presents the results. The summary table included in the figure illustrates the results of the ANOVA and Tukey's HSD results with same color shading for the same statistical groups and no shading if a particular instance was statistically equivalent to other climates. From the results, it is apparent that SN was significantly lower in the DW climatic region, while WC had the highest SNs. The other climatic regions had values in between these two extremes.

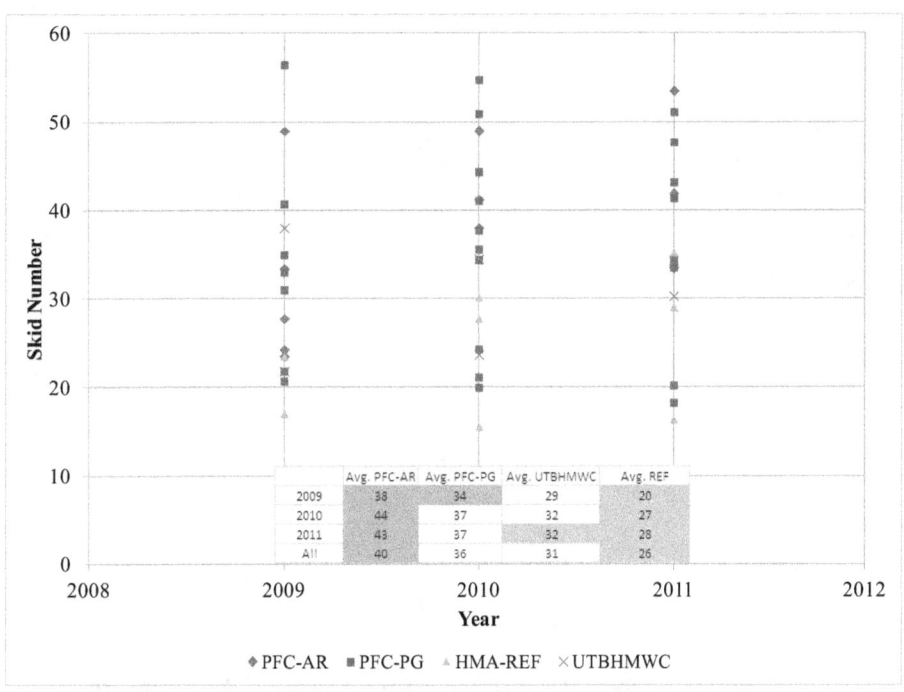

Figure 52. Effect of Binder/Mixture Type on SN.

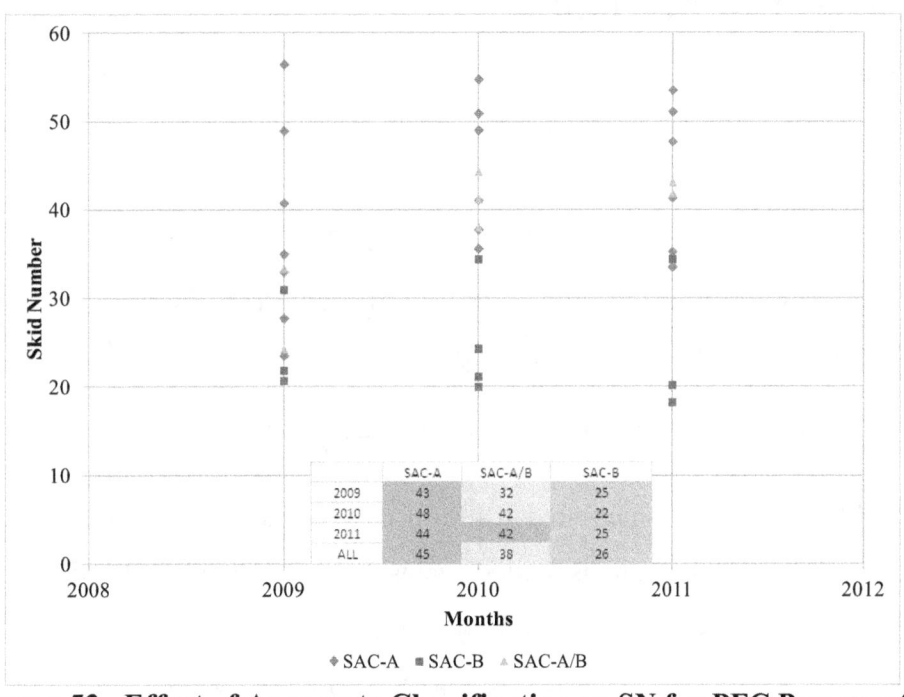

Figure 53. Effect of Aggregate Classification on SN for PFC Pavements.

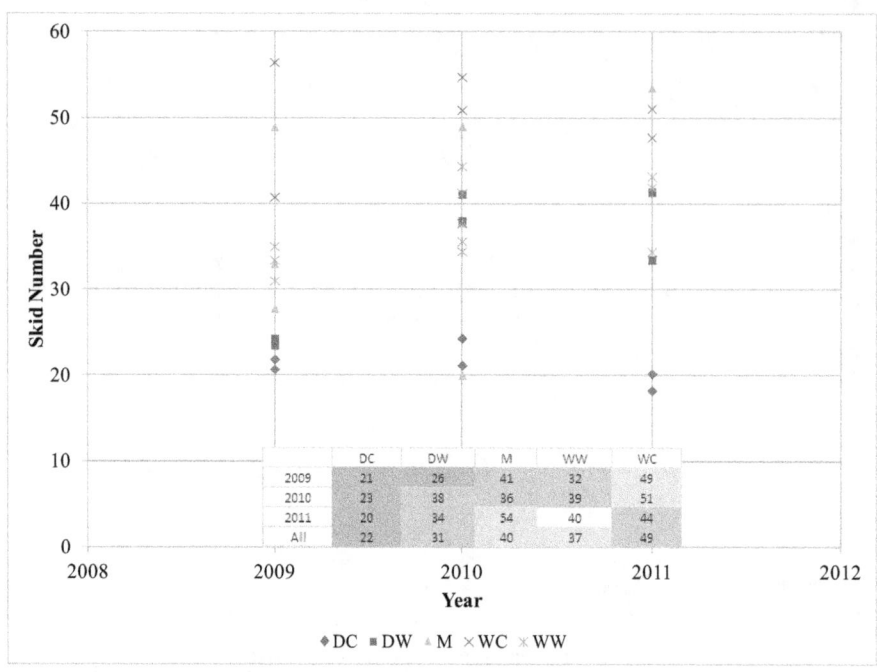

Figure 54. Effect of Climate on SN for PFC Pavements.

INTERNATIONAL FRICTION INDEX

Following the procedure described in ASTM E1960, the texture and friction values were used to calculate the International Friction Index (IFI). First, the speed constant (S_p) is determined using the MPD as follows:

$$S_p = 14.2 + 89.7 MPD$$

Next, the friction measured at slip speed is used to adjust the friction to a common slip speed of 60 kph using the following relationship:

$$FR60 = FRS \times EXP\left[\frac{(S-60)}{S_p}\right]$$

where *FRS* is the friction measured at slip speed and *S* is the slip speed (i.e., 20 kph). Then the friction measured at 60 kph is plotted against the calculated FR60 and a regression line adjusted to the data to obtain the calibrated friction number, *F60*:

$$F60 = A + B \times FR60$$

IFI is reported in terms of *F60* and *Sp*. Figure 55 illustrated the relationship between *F60* and skid number. A good correlation between these two values exists, especially for AR-PFCs. The regression equation for each case is shown, and the R-squared for all PFC pavement sections was 0.81 (see Figure 55c). The IFI threshold corresponding to a SN of 20 is 0.2, and 0.3 for a SN of 30.

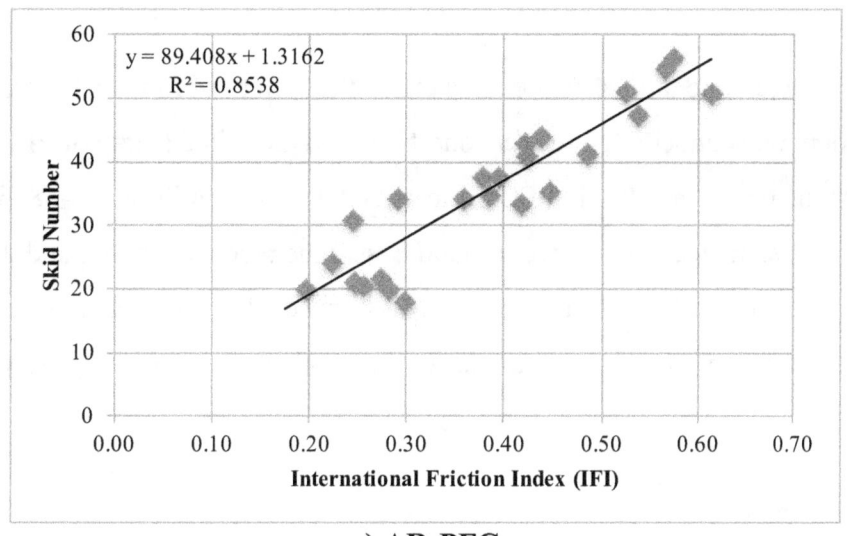

a) AR-PFC

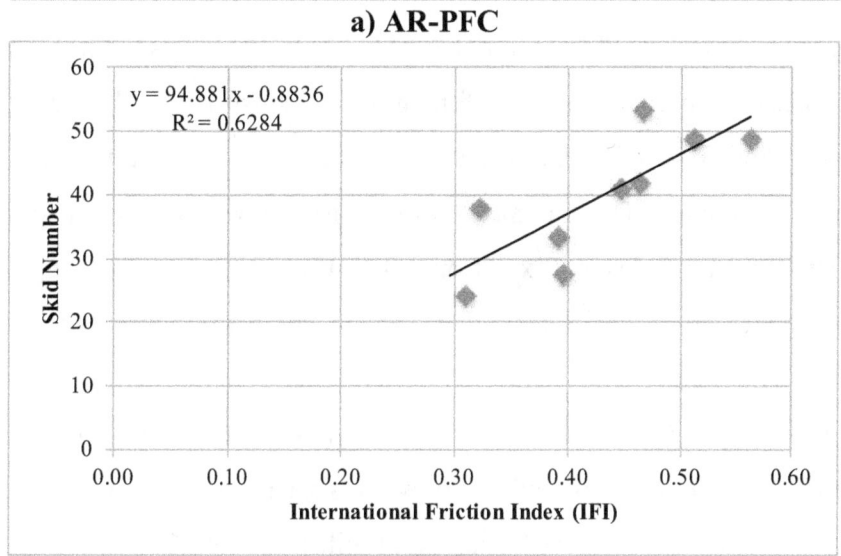

b) PG-PFC

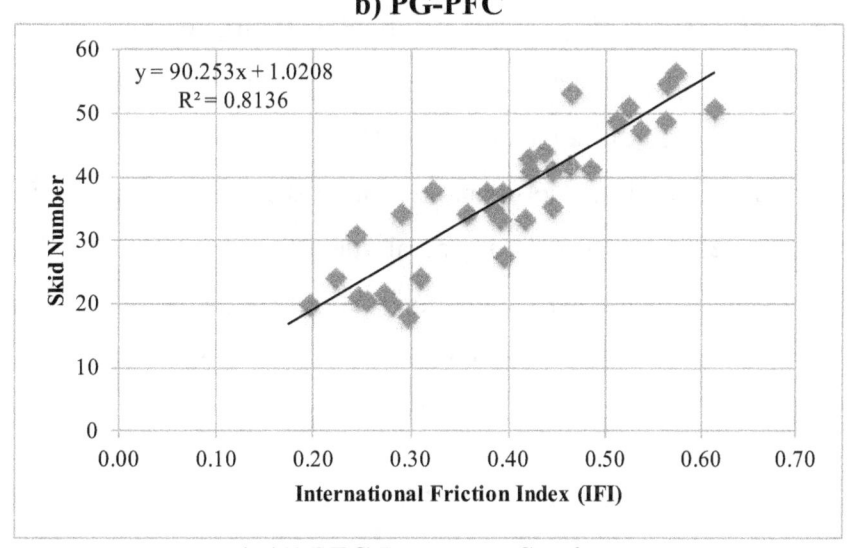

c) All PFC Pavement Sections

Figure 55. Relationship between SN and IFI.

ACCIDENTS

Analyzes were done to investigate the wet weather safety of PFC in Texas, specifically looking at accident rates under dry and wet conditions and the initial safety of PFC following construction. At the onset, the 20-odd PFC projects under review in Project 0-5836 were evaluated. Since no significant trends were found, researchers decided to expand the dataset to include a larger sample of PFC constructed in Texas between 2003 and 2011. A total of 601 PFC projects were identified, providing a better platform to evaluate wet weather safety of PFC across Texas.

PFC Projects in Texas

To evaluate the safety of PFCs, a number of projects using these mixes were identified and the crash statistics on these compared before and after construction. The University of Texas at Austin maintains a database of HMA projects in Texas—the Texas Cartographic Information Technology (TxCIT) database. This database was used to identify PFC projects; 161 PFC projects constructed as main lanes in Texas from 2003 were identified. This sample does not include PFC constructed on service or frontage roads, on and off ramps or connectors, and flyovers.

The lengths of the projects selected for the analyses ranged from 1 to 20 miles, and included both undivided and divided roads with two to six lanes per direction. The average lane miles for all the PFC projects considered is about 24 miles. Figure 56 shows the relative breakdown of these PFC projects by highway facility. PFC is targeted for high-volume interstate highways (IH) and is generally only used on higher volume farm- or ranch-to-market (FM/RM) roads as well as national (US) and state highways (SH).

The projects identified for the crash analyses are distributed throughout Texas and span the five climatic zones (see Figure 57). As a general rule, PFC is no longer used in very cold/wet regions where the possibility of black-ice conditions severely impacts the safety and performance of these mixes. Figure 57 shows that PFC is favored in wet/warm climates.

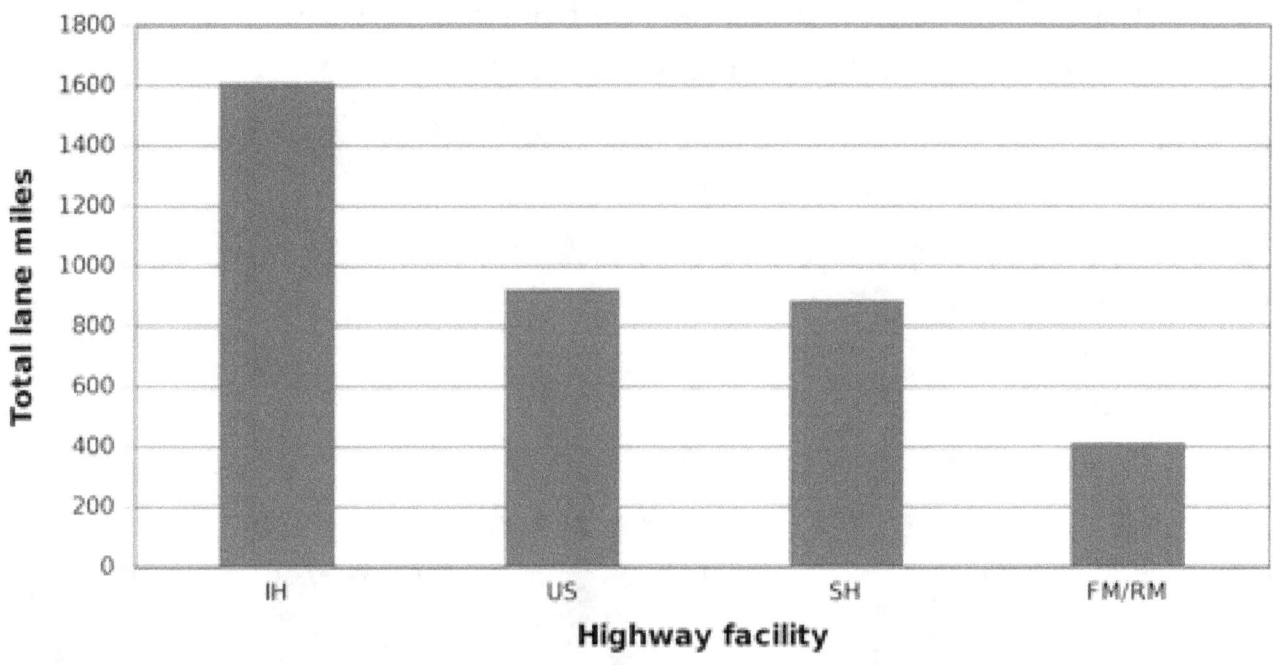

Figure 56. Breakdown of PFC Projects by Highway Facility.

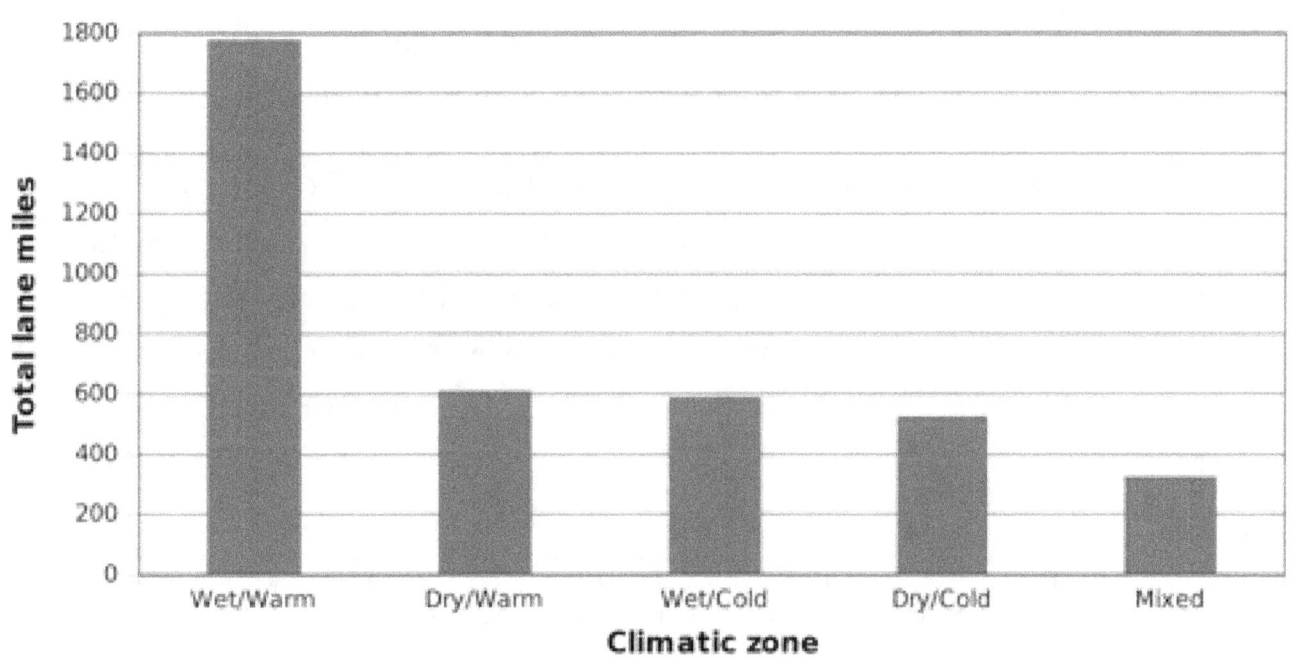

Figure 57. Breakdown of PFC Projects by Climatic Region.

Figure 58 shows the breakdown of PFC projects by population area. The majority of PFC is constructed in rural areas (< 5k people), but includes sections in small (5k–50k), large (50k–200k), and urbanized (> 200k) regions. This breakdown is relevant when evaluating the relative safety of PFC compared to other HMA types since the vehicle miles traveled (VMT) and exposure density is generally larger in urbanized areas.

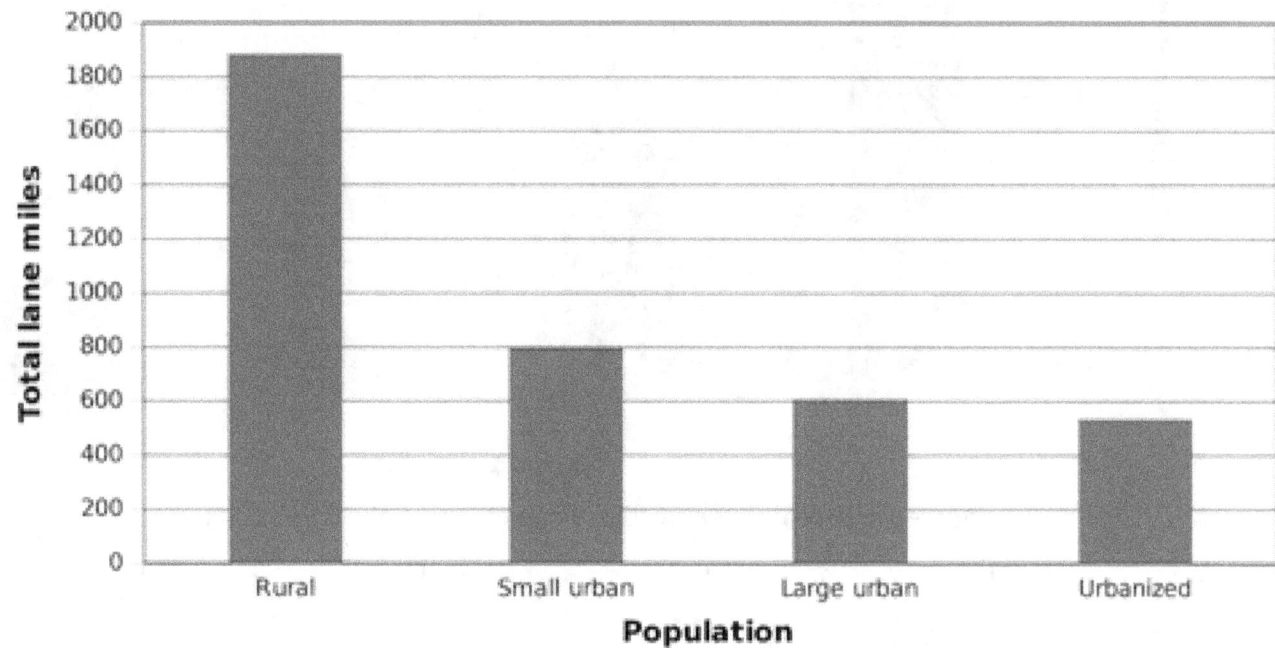

Figure 58. Breakdown of PFC Projects by Population.

PFC in Texas is manufactured using either polymer modified PG binder or AR binder. There were three times as many PFC projects with PG binders in the selected sample. The objective of the analysis was to evaluate the safety of PFC in Texas, specifically to identify the benefits of PFC under wet conditions. In addition, the research comments on the safety of PFC immediately following construction when skid resistance may be compromised.

Crash Database

TxDOT CRIS database was used to obtain accident data. The locations of accidents are reported in geographical coordinates and in displacements from Texas reference marker locations. The latter coordinate system was referenced against the beginning and ending reference markers for construction projects, allowing the accidents in CRIS to be related to the surface mixes used on these construction projects, as archived in TxCIT.

Figure 59 shows the annual total number of accidents reported in Texas on all roads compared with the annual statewide rainfall totals obtained from National Oceanic and Atmospheric Administration (*33*). A decreasing trend in the number of annual accidents is evident after 2007. The sudden and continuing decrease in accidents is in line with national statistics that the National Highway Traffic Safety Administration (*34*) reported and is related to the slowing economy and the subsequent drop in vehicle miles traveled (VMT). Although there appears to be a relationship between observed rainfall and accidents, the trend is rather weak when considering the many other factors influencing the causes of road accidents.

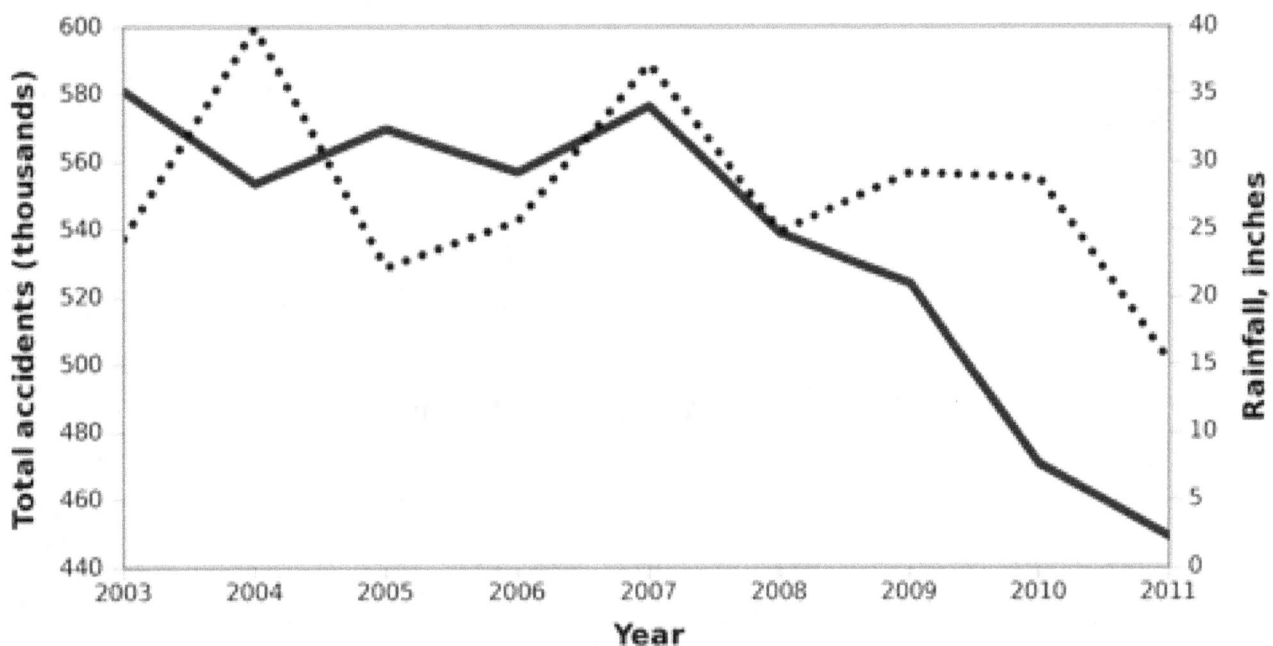

Figure 59. Accident and Rainfall Data for Texas.

For the 161 PFC project locations identified, a total of 39,712 accidents were reported between 2003 and 2011. Figure 60 shows the number of reported accidents on the highway facilities between 2003 and 2011. The proportion of accidents on the various highway facilities matches the respective lane miles of project sections on these facilities. While the number of accidents fluctuate from year to year, a decreasing trend post-2007 in the number of accidents is evident, which coincides with the general decreasing trends observed throughout Texas. A similar breakdown in number of accidents was observed in the different climatic zones, as shown in Figure 61. While the number of reported accidents in rural areas is greater given the larger number of projects in these areas, the number of accidents in small urban areas is less compared to large urban areas, which again may be related to the greater VMT with higher population density.

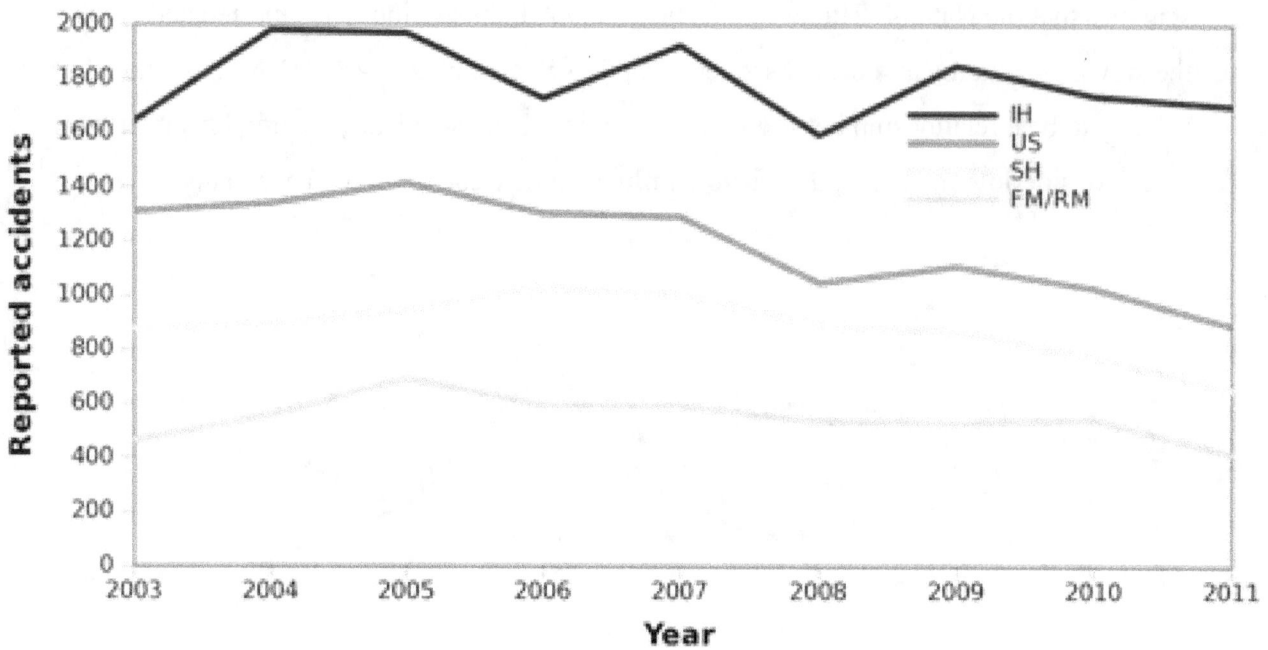

Figure 60. PFC Accidents on Highway Facilities.

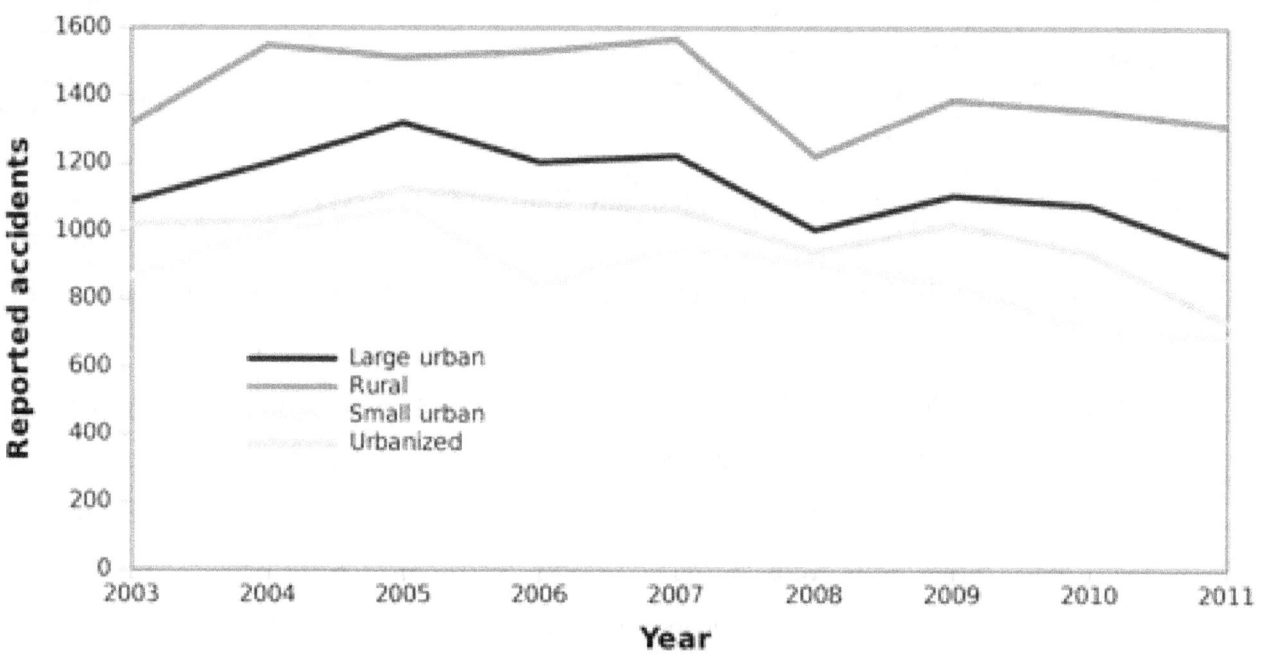

Figure 61. PFC Accidents by Population Zone.

Similar trends were observed for the total number of injuries and fatalities reported by facility type, climatic zone, and population density, which are expanded on in more detail in the following section.

Crash Analysis

The previous section provided overall breakdowns of accidents reported on the identified sections on a yearly basis between 2003 and 2011. PFC was constructed on these sections sometime between 2003 and 2011. To provide an accurate statistic of PFC safety, it is necessary to know exactly when PFC was constructed and specifically the end-of-construction dates for these projects. TxCIT includes an acceptance date, which is the date when TxDOT accepts the project from the contractor. This date was used in the analyses and assumed to correspond to the date when the road is reopened to the public. The majority of the PFC projects identified are rehabilitation overlay projects typically used to replace old dense-graded Type-C (TYC) projects at the end of their service lives. Of the 39,712 accidents reported, about one-third of these (13,405) occurred when the project included a PFC surface, i.e., after rehabilitation of the existing TYC.

CRIS reports the weather condition at the time of the accident. Of the accidents reported, 84 percent occurred under clear or cloudy conditions and 14 percent occurred in rainy conditions with the balance occurring under windy, snowy, sleet, hail, or unknown conditions. CRIS also reports the road condition at the time of the accident, i.e., whether wet or dry. Of all the accidents, 78 percent occurred when the road was reportedly dry and 19 percent when wet. Figure 62 shows a breakdown of the accidents on the projects by weather and road condition with the relative proportion of the accidents that occurred under these conditions when the surface was a PFC and prior to rehabilitation (TYC), providing a general indication that accidents on PFC are comparatively lower under wet conditions. The analyses showed no significant difference in the proportion of accidents on PFC and TYC as influenced by lighting (light, dark, dawn, or dusk) or road alignment at the accident site (straight or curved sections).

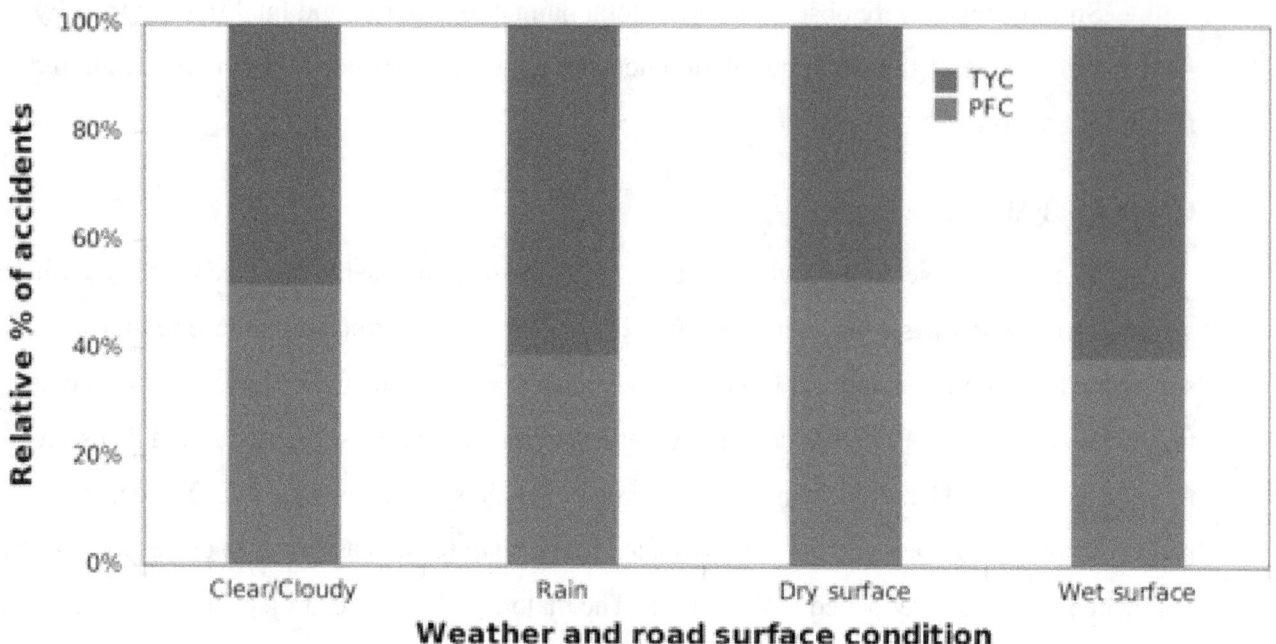

Figure 62. Relative Breakdown of Accidents by Weather and Road Condition.

Accident Rates

While indications are that PFC appears to reduce the number of accidents occurring under wet conditions, the CRIS statistics are not conclusive. As indicated previously, there is a disparity in the percentage of accidents that occurred under reported rain conditions and accidents on wet roads, indicating that the road condition as reported in CRIS does not necessarily reflect the weather condition at the time of the accident, or vice versa. An analysis of the accident data indicated that 6 percent of the accidents reported occurred when the road was wet under clear or cloudy conditions. Furthermore, no clear difference in the accident rates observed before or after construction of PFC was found when comparing statistics on accidents occurring on wet surface as CRIS reported. Figure 63 illustrates the cumulative number of accidents on the PFC sections per lane mile. The x-axis indicates the time in months before and after construction of the PFC sections. The analysis was done by adding the accidents per lane mile 72 months before and 72 months after construction for all the PFC sections. No distinct difference in the accident rates before and after construction of PFC is observed when the road was reportedly wet. A trend line of the pre-construction accidents has been superimposed. Based on the derivative of the data in Figure 63, the annual accident rate on the sections before construction of PFC is 0.46 accidents per lane mile and remains the same after construction. For similar analyses including all data under all conditions, the pre-construction rates remain at 0.46

accidents per lane mile while the post-construction rates reduce slightly to 0.43 accidents per lane mile. This was also the case for the projects when the road surface was reported dry.

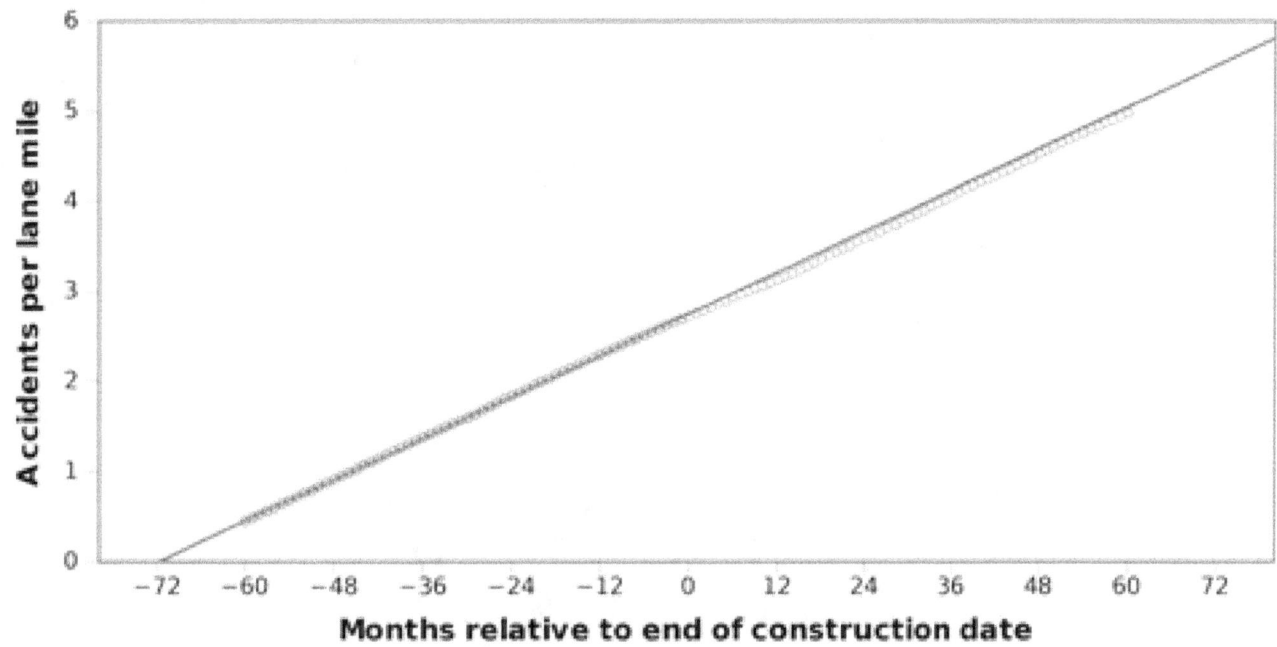

Figure 63. Accident Rates of CRIS Reported Wet PFC Sections.

To evaluate more accurately the possible effect of rain of PFC safety, rainfall data collected at 280 weather stations throughout Texas was used, as archived in the Global Surface Summary of Day (GSOD) database that the National Oceanic and Atmospheric Administration maintained (*35*). This database includes daily rainfall totals reported at mainly municipal airport facilities across Texas. Daily rainfall totals at these weather stations between 2003 and 2011 were used to indicate rain conditions on the dates and in the vicinity of the accidents reported in CRIS. Figure 64 shows locations of the PFC accidents and GSOD weather stations in Texas.

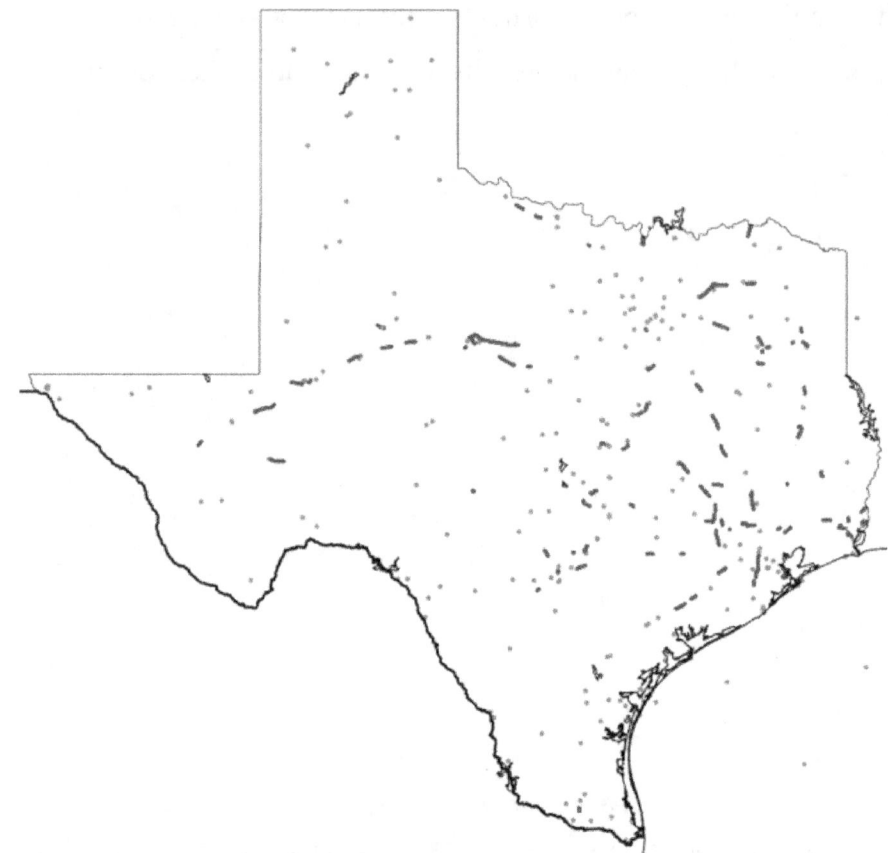

Figure 64. PFC Accidents and Weather Stations in Texas.

Of all the accidents reported, 36 percent occurred within 5 miles of a GSOD weather station, 71 percent within 10 miles, 83 percent within 15 miles, and 90 percent within 20 miles of a weather station. By interpolating the daily rainfall totals from the weather stations in the vicinity of the crash locations, it is possible to estimate the rainfall that occurred on the day of an accident. Obviously, given the spread and nature of rainfall patterns, the closer the location to the weather station, the more accurate the prediction. Using this approach, accidents that occurred when the daily rainfall in the vicinity of the accident exceeded 0.1 inches were identified.

Figure 65 shows the accident rates for PFC sections having estimated daily rainfall totals exceeding 0.1 inches. These are based on accidents that occurred within 5 miles of a weather station. From these data, the annual accident rate for the sections prior to PFC construction was 0.47 accidents per lane mile; for the PFC sections, this is reduced to 0.44 accidents per lane mile, a marked reduction. Similar analyses were done for accidents that occurred within 10, 15, and 20 miles of a weather station. Table 8 summarizes the accident rates determined for these

showing the rates in accidents per lane mile occurring within the stated distances or radii from a weather station.

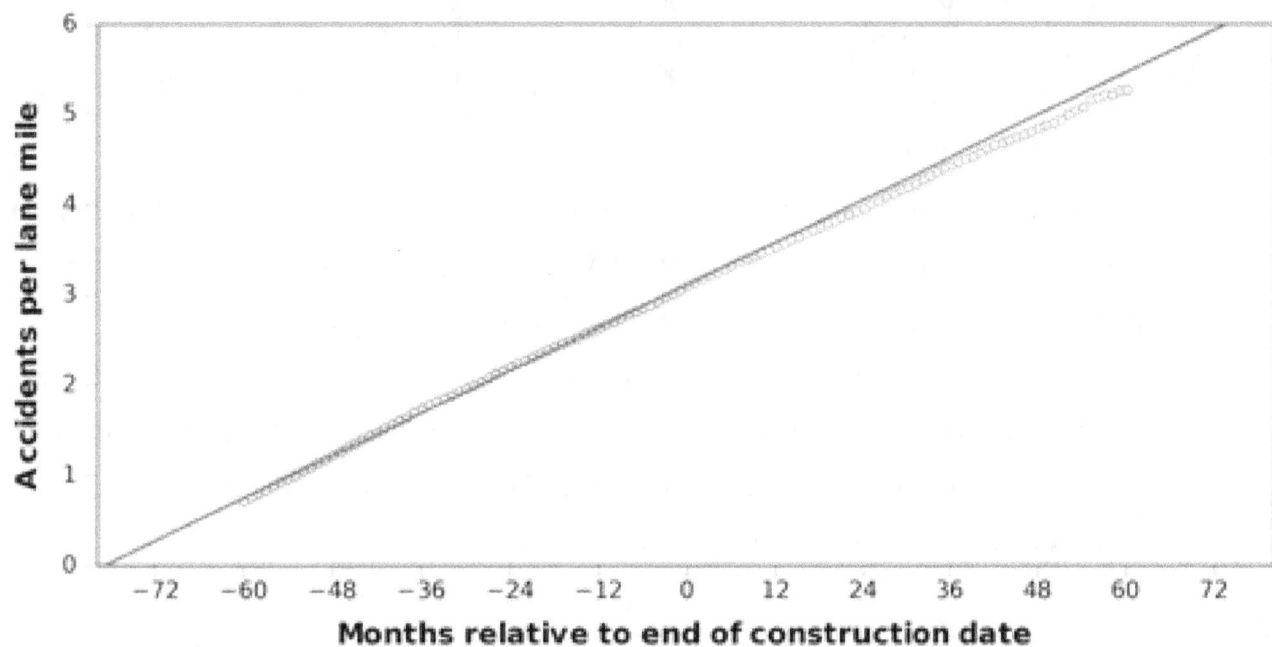

Figure 65. Accident Rates of PFC Sections with Daily Rainfall Exceeding 0.1 Inch.

Table 8. Annual Accident Rates of Analysis Sections.

Radius, miles	TYC	PFC
5	0.4724	0.4411
10	0.4805	0.4258
15	0.4848	0.4478
20	0.4863	0.4483

Similar trends in accident rates per lane mile were observed when evaluating injuries and fatalities on the PFC sections when rainfall exceeded 2.5 mm (0.1 inch). Table 9 shows these rates for accidents occurring within the specified radii of weather stations used to determine if the rainfall totals exceeded 2.5 mm (0.1 inch) on the day of the accident. Given the low number of fatalities and the lower number of accidents that occurred within 5 miles of a weather station, the fatality rates for these occurrences are less accurate but appear to stabilize as more accidents are included in the analyses. Nonetheless, a marked decrease in accident, injury, and fatality rates are evident for the PFC sections.

Table 9. Annual Injury and Fatality Rates of Analysis Sections.

Radius, miles	TYC		PFC	
	Injuries	Fatalities	Injuries	Fatalities
5	0.2504	0.0058	0.2207	0.0015
10	0.2507	0.0043	0.1997	0.0037
15	0.2639	0.0059	0.2116	0.0038
20	0.2653	0.0061	0.2092	0.0046

Safety of PFC Following Construction

To investigate the reported higher accidents on PFC immediately following construction, a plot of all the accidents per lane mile on the PFC sections was made (see Figure 66). Some research studies have reported similar observations claiming that apparently drivers take advantage of the reduced splash and spray on PFCs during rain events, driving faster at shorter following distances and causing more accidents (*36,37*). Figure 66 shows the accidents per lane mile six months before and after construction of the PFC sections regardless of weather or surface condition, i.e., considering all accidents reported in CRIS.

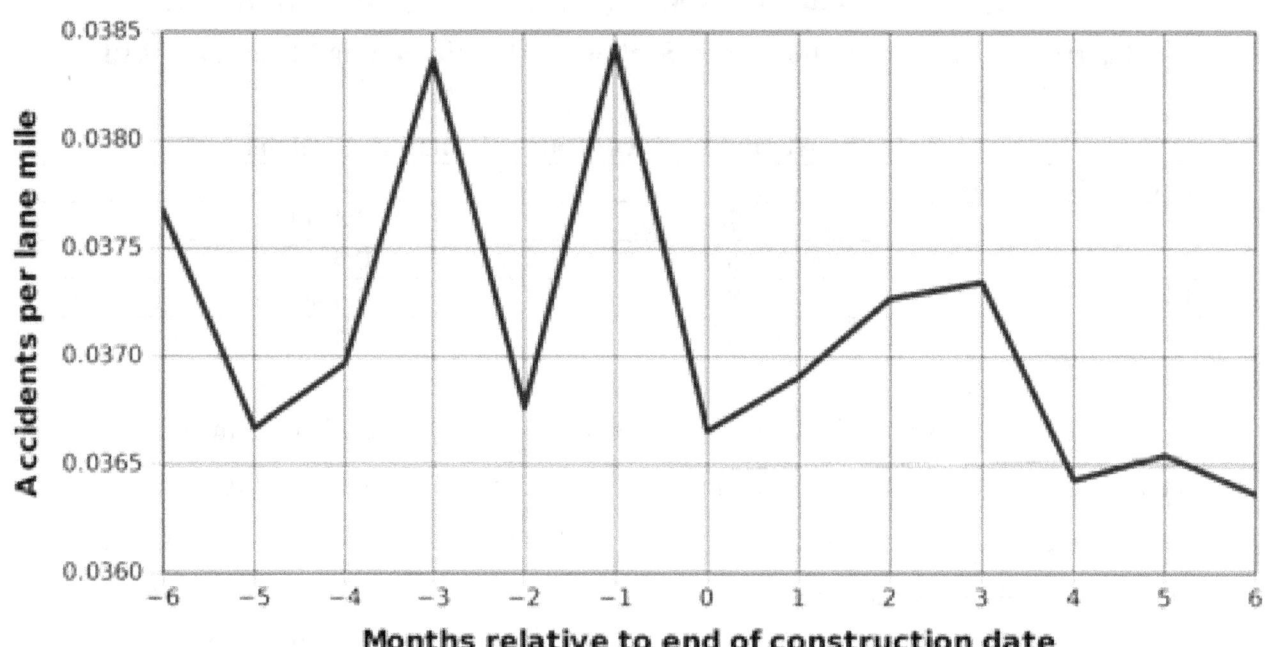

Figure 66. Accidents per Lane Mile on PFC Sections under Varying Conditions.

Figure 66 shows a definite increase in the number of accidents that continues for three months following construction of PFC. The increase in accidents is slight, however, in the order of about 0.0005 accident per lane mile. Nonetheless, TxDOT should investigate measures to improve this initial safety concern, perhaps by scarifying the road surface or by applying sandblasting or any other similar technique that improves texture right after construction. The same recommendation applies to all hot-mix asphalt surfaces, not only PFCs.

SERVICE LIFE

Several pavement sections were included year-to-year in the experimental design when performance issues were reported. Besides the on-site measurements, field cores were acquired and forensic studies performed to determine the possible causes of the observed distress. A description of the issues reported in these sections and forensic study results is included next. Table 1 lists additional details of these sections.

Section 19, IH 20, Tyler

This pavement section was located on IH 20, north of the town of Canton in Van Zandt County, Tyler District. The overall performance was adequate, with WFVs between 56 and 63 seconds OWP, and between 39 and 54 seconds BWP, and adequate texture and friction. However, in three small areas of the eastbound lanes, excessive bleeding and rutting was observed. The areas or patches were about 1.8 m (6 ft) long by 0.9 m (3 ft) wide. Two of the patches were located on the inside lane, while the third and largest patch spanned from the inside lane over into the outside lane's left wheel path (see Figure 67a). Figure 67b shows that rutting occurred within the areas of bleeding, and a slight ridge formed along the centerline where the softened mixture shoved away from the wheel path.

a) Overview of the Bleeding Patches	b) Detailed View of Bleeding and Rutting
Figure 67. Pavement Section on IH 20 in Tyler.

Several cores were obtained from the site for forensic testing to determine the possible cause of the distress. In the laboratory, researchers noted a noticeable heavy smell of either gasoline or diesel fuel coming from the cores. A fuel leak at the location could explain the smell and observed bleeding of the binder. When a solvent like gasoline leaks onto asphalt pavements, it affects the asphalt binder, making it softer. Over time, gravity and repeated traffic loads cause the denser aggregates in the pavement to settle, leaving the soft binder on the surface of the pavement. In the case of PFCs, the highly connected AV structure could allow solvents to easily permeate the pavement structure. It is likely that an accident and subsequent fuel leak occurred at that location.

Section 26, US 281, Pharr

This pavement section was located on US 281 in Edinburg, Hidalgo County, within the Pharr District. The test section spanned from SH 495 to Trenton Road. The PFC pavement was constructed in year 2004 using 9.1 percent AC-10 + 17.7 percent AR binder, gravel aggregate (SAC-A), and 1 percent lime per mix design. In 2010, severe raveling was apparent in certain areas of the pavement section. An area of the test section was in good condition as compared to another area that the distress had affected. That year, on-site WFV and cores were obtained from the pavement. Texture and friction measurement were not performed due to conflicting scheduling of the equipment. Fourteen cores were extracted for completing a forensic analysis; seven cores BWP in the good performing area, and seven OWP in the area with raveling. Figure 68 shows pictures of the areas where the cores where acquired. Out of the seven OWP cores, two were damaged while grinding and thus only five were available for the forensic tests.

    a) Area in Good Condition   b) Area with Raveling Distress

Figure 68. Pavement Section on US 281 in Pharr.

Dimensional analysis, CoreLok®, permeability, as well as binder content and aggregate gradation after extraction were performed on the field cores. Table 10 lists the results of these tests and the on-site WFVs.

Table 10. Forensic Test Results for US 281 in Pharr.

Location	WFV (sec)	Dimensional Analysis AV (%)	CoreLok® AV (%)	CoreLok® Porosity (%)	Permeability (m/day)	Binder Content (%)
Raveled Area	34 (OWP)	27.4	24.8	23.5	108.7	4.5
Good Performing Area	62 (BWP)	19.0	19.3	18.9	49.4	5.7

The on-site WFVs were about a half-minute on the raveled area OWP and up to a minute in the good performing area BWP, which is acceptable for PFCs. The dimensional analysis and CoreLok® procedures yielded similar results; AV in the raveled area were larger than the AV measured using the cores from the good performing area. The percent porosity was adequate, between 95–97 percent of the total AV. Similar to the on-site WFV measurements, significant changes in the laboratory-measured permeability were observed, with more than double the value obtained using the cores from the raveled area vs. the cores obtained from the good performing area.

The binder content was calculated after extraction, assuming all the rubber was extracted. However, since rubber is not soluble on trichloroethylene, some rubber particles remained attached to the aggregates after the extraction procedure was completed. Assuming no rubber was extracted in the process, the binder contents would be 5.3 and 6.7 percent for the raveled and good performing areas, respectively. These values are still much lower than the required binder content for AR-PFCs (i.e., 8–10 percent).

Following the extraction procedure, the aggregates were sieved to determine the actual aggregate gradation used in the field. Results are shown in Figure 69. A note in the mix design clarified that the district lab allowed the use of the master gradation corresponding to PG-PFC mixtures. The mix design gradation was outside of these limits for sieve sizes 1/2 inch, No. 4, and No. 8. The aggregate gradation of the good performing area followed closely the mix design. The raveled area had a finer aggregate gradation, yet, except for sieve size 1/2 inch, it fell between the boundaries of the gradation limits.

X-ray computed tomography (CT) and image analysis techniques revealed the AV distribution as well as the interconnected AV content of two of the cores taken from the site (Figure 70). The CoreLok® AV content was used to calibrate the image analysis. The interconnected AV resulting from the image analysis overlapped with the total AV content curves, demonstrating no loss in AV interconnectivity.

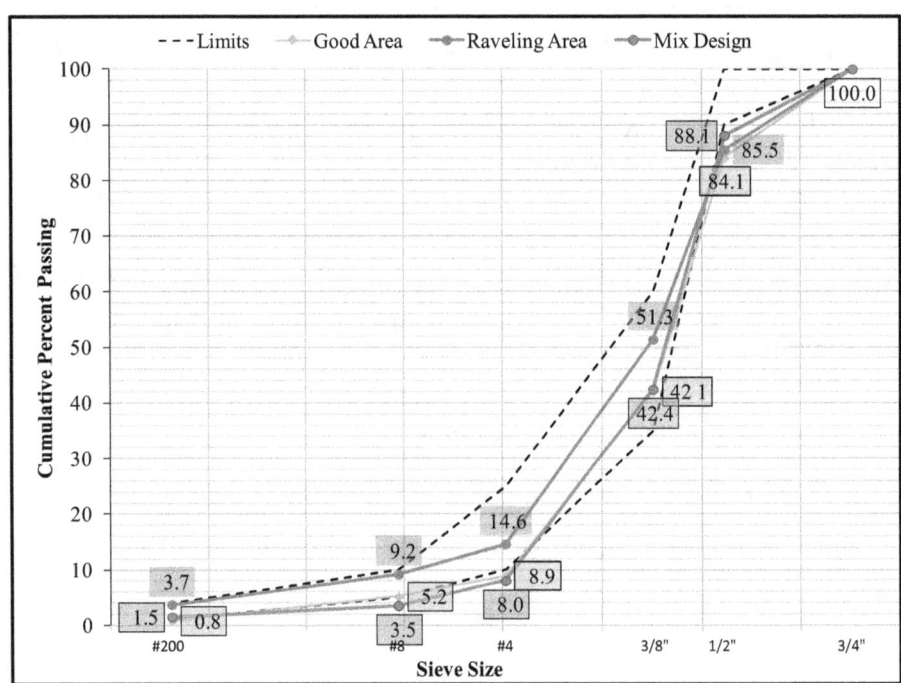

Figure 69. Aggregate Gradation after Binder Extraction for US 281 in Pharr.

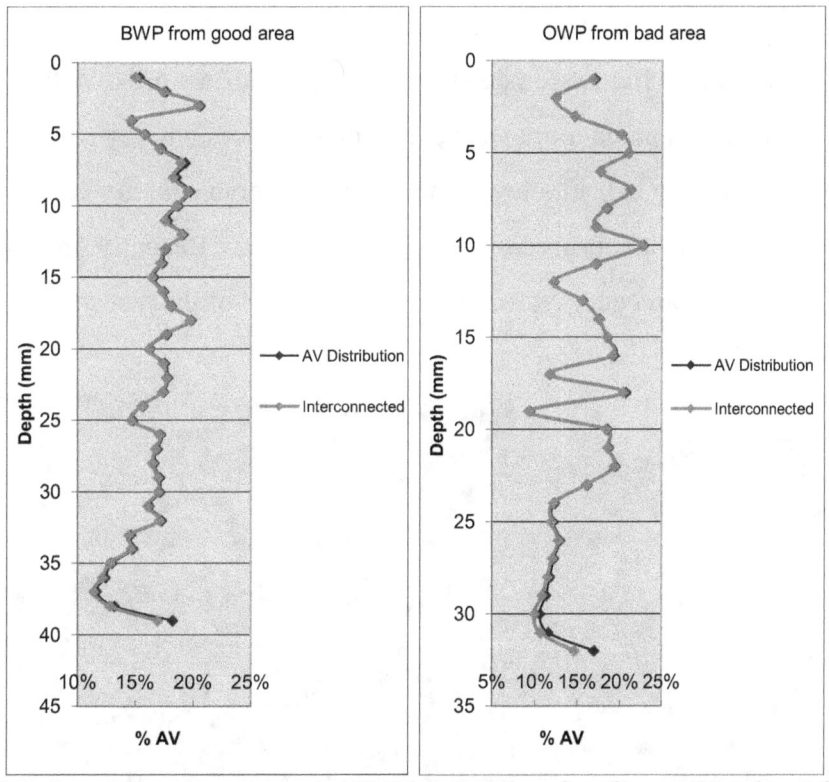

Figure 70. Total and Interconnected AV Distribution for Cores Obtained from US 281.

The cause for the observed raveling is possibly due to the low asphalt binder content in the mixture, combined with poor construction practices in the area of the pavement section that showed this distress. Given that some portions of the PFC pavement section were in good condition, it is also possible that the performance issues were caused by asphalt draindown during transportation of the mixture from the plant in Harlington, which was around 35 miles away. During a hot day in South Texas and over that distance, it is plausible for draindown to occur. In addition, binder aging could explain some of the observed asphalt-aggregate separation (especially from the small river gravel aggregate) and the tendency of the field cores to easily lose aggregates and crumble.

Section 27, US 90, Houston

This pavement section on US 90 was located just west of Brookshire in Waller County, within the Houston District. The PFC pavement was constructed in 2004 using 8.3 percent AC-10 + 18 percent AR binder and sandstone aggregate (SAC-A) per mix design. The test section began at westbound mile marker 794-0.276 and continued for 610 m (2,000 ft), extending back east into the center of the town of Brookshire. This pavement section was visited in measurement year 2010 after reports of excessive raveling (Figure 71). The performance of this pavement was adequate for about five years, but started deteriorating rapidly afterwards. Within the town of Brookshire, the raveling was much more severe, especially places where vehicles turned to access businesses. Continuing east into town, the crumb rubber eventually gave way to another PG-PFC pavement section that appeared to be in good condition. Recently, the AR-PFC was overlaid from the PG-PFC boundary, west, to a point just short of the test site for this project.

Figure 71. Raveling Distress on US 90 in Houston.

The WFVs collected on-site showed that the pavement was impervious, with flow times greater than 5 minutes, which was set as the field threshold. Texture measured in the field resulted in MPDs of 1.7 mm (0.067 inch) BWP and 1.4 mm (0.055 inch) OWP, which are comparable to other sites. Friction measurements, on the other hand, were higher than the typical values for other pavement sections, resulting in 0.64 BWP and 0.60 OWP. It is common to observe an increase in friction in raveled pavements. The skid number measured in year 2009 was about 24, which is low.

Two cores were obtained from this section to conduct forensic tests and determine the possible causes of raveling and poor drainability. The laboratory permeability test confirmed the findings from the WFV test performed in the field, which is that the section is essentially impervious for both OWP and BWP. The average Rice specific gravity of the cores was 2.311, and the individual sample values were within the standard deviation limits. CoreLok® tests on the cores yielded a bulk specific gravity of 2.045, an average AV of 11.5 percent, and an average percent porosity of 10.9 percent.

Ignition oven test were also performed to estimate the binder content in the mixture. The average binder content was 9.1 percent, which is typical for AR-PFCs, although the design content was 8.3 percent. Following the ignition oven procedure, the aggregates were sieved to determine their actual gradation. Figure 72 shows the results. All cores had similar aggregate gradations that were much finer than the mix design and above the upper specification limit for sieve sizes 9.5 mm (0.38 inch), No. 8, and No. 200. It is relevant to note that the limits did not follow TxDOT specifications for either AR-PFCs or PG-PFCs, but were noted as shown on plans.

Application of X-ray CT and image analysis techniques revealed the AV distribution as well as the interconnected AV content of the road cores. Results are presented in Figure 73. For the X-ray CT analysis, images were acquired at intervals of 1 mm (0.039 inch) in the vertical direction. The CoreLok® AV content was used to calibrate the image analysis and generate the profiles in Figure 73. The average interconnected AV resulting from the image analysis was 10.2 percent for Core 1 and 10.7 percent for Core 2. These values are similar to the percent porosity value of 10.9 percent resulting from the CoreLok® test. The interconnected AV was extremely low for PFCs, which usually have over 95 percent AV interconnectivity.

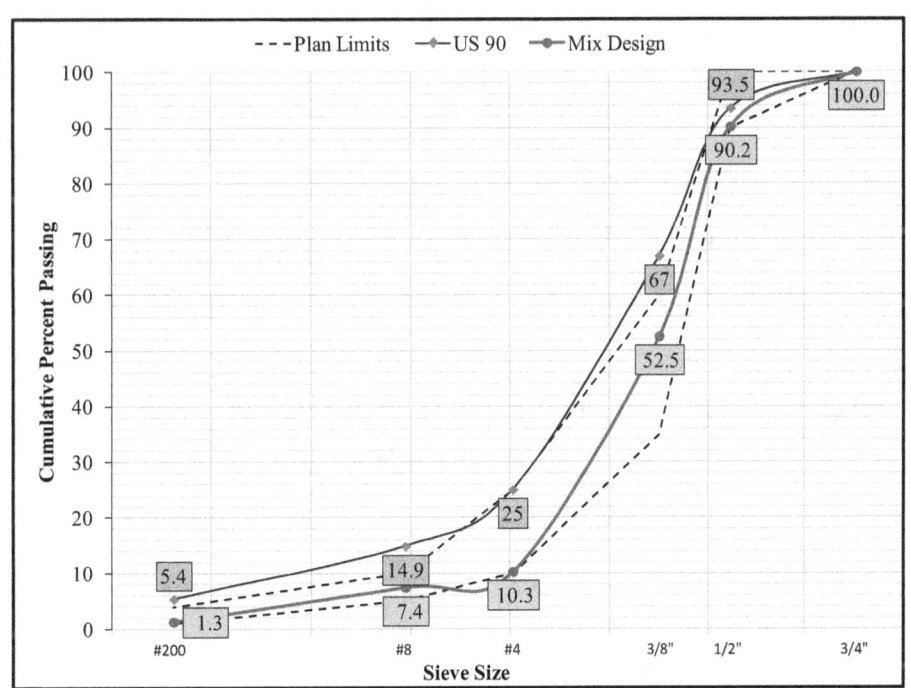

Figure 72. Aggregate Gradation after Ignition Oven for US 90 in Houston.

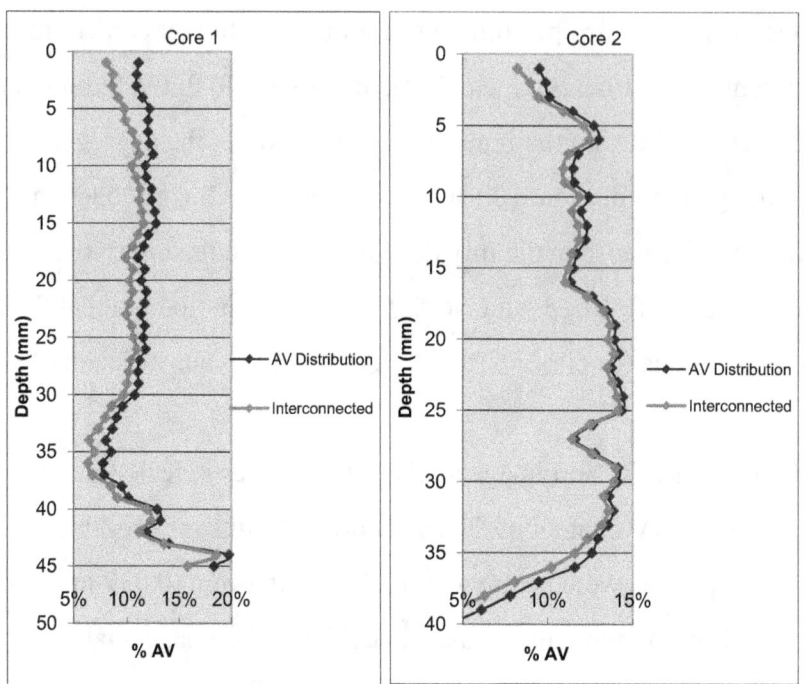

Figure 73. Total and Interconnected AV Distribution for Cores Obtained from US 90.

From Figure 73, it is apparent that near the surface of the cores is where the loss of pore interconnectivity was most severe (especially for Core 1), with the level of severity steadily decreasing with depth. Ideally, the total and interconnected AV should be homogeneous and very similar to each other. In Figure 73, this would translate into practically straight and overlapping AV distribution and interconnected lines at the designed AV content value of around 18 to 22 percent.

The cause of reduced drainability for this AR-PFC pavement is likely related to the aggregate gradation and asphalt binder content. Figure 73 shows that the actual mixture gradation was finer than both the design and plan specification limits. The binder content was also on the upper threshold for PFCs. With regard to the observed raveling, a contributing factor could be the amount and type of traffic on this section. The researchers observed during testing and coring that a high portion of the traffic was comprised of trucks traveling on US 90 to get to I-10. Further into the town of Brookshire, the raveling was most severe at the entrances to industry locations where heavy trucks were the primary mode of hauling materials.

Section 28, SL 289 FR, Lubbock

This pavement section on SL 289 frontage road was located in Lubbock County. The section extended from FM 1730 (Slide Road) to IH 27. Per mix design, this mixture included

8.6 percent AR + 18 percent AR. Construction took place late in 2010. Within a month of construction, the section experienced severe raveling that was more prominent at intersections and turn lanes (see Figure 74).

Figure 74. Raveling Distress on SL 289 in Lubbock.

WFVs were measured in year 2011 both OWP and BWP. Results listed in Table 11 indicate that the right lane had lower WFVs as compared to the left lane. The WFVs were also higher OWP than BWP. Higher WFVs correlated with the areas with more severe raveling. Possible contributing factors to the observed distress include:

- Construction late in the season at low ambient temperature.
- Use of AR binder.
- Lack tack coat under the PFC pavement.
- Lack of AR binder design or QC/QA.
- Section opened too early to traffic.

In addition to the on-site measurements, the research team obtained two field cores: one OWP around the area where the WFV was 90 seconds and another BWP around the area where the measured WFV equaled 16 seconds. Forensic laboratory tests performed on the cores included dimensional analysis, CoreLok®, permeability, as well as binder content and aggregate gradation after extraction. Table 11 lists the results of these tests. Binder content was calculated after extraction, assuming all the rubber was extracted with the binder. The values ranged from 7.3 to 7.9 percent BWP and OWP, respectively, which are about 1 percent lower than the target binder content of 8.6 percent specified in the mix design. The AV content of the field cores was close to the typical range for PFCs.

Table 11. Forensic Test Results for SL 289 FR in Lubbock.

Location	WFV (sec) [a]	Dimensional Analysis AV (%)	CoreLok® AV (%)	CoreLok® Porosity (%)	Permeability (m/day)	Binder Content (%)
OWP	21 (RL,RWP) 21 (RL,LWP) 90 (LL,RWP) 41 (LL,LWP)	23.8	19.2	18.6	- [b]	7.9
BWP	16 (RL) 38 (LL)	24.4	21.1	19.6	63.6	7.3

[a] RL = right lane; LL = left lane; RWP = right wheel path; LWP = left wheel path
[b] OWP core had insufficient thickness to perform the permeability test

After binder extraction, the aggregates were sieved to determine the actual mixture gradation. Results are illustrated in Figure 75. The gradations were finer when compared to the design aggregate gradation and above the specification limit for sieve sizes No. 4 and No. 8. The aggregate gradation for the OWP core had even finer gradation than the one obtained BWP. This explains the higher WFVs measured in that area (i.e., 90 seconds).

Application of X-ray CT and image analysis techniques were used to quantify the AV distribution and the interconnected AV content of the road cores. Results are presented in Figure 76. For the X-ray CT analysis, researchers acquired images at intervals of 1 mm (0.039 inch) in the vertical direction. The CoreLok® AV content was used to calibrate the image analysis and generate the profiles in Figure 76. The total and interconnected AV profiles overlap, which is consistent with the percent porosity CoreLok® results. The total AV at the surface of the OWP Core is somewhat lower than the BWP core (i.e., less than 40 percent vs. almost 50 percent). In addition, the interconnected AV profiles diverge in the case of the OWP core with depth.

The cause of the observed raveling in this section was likely due to the time of the year the pavement was constructed combined with poor construction practices. In addition, the lower binder content in the mixture was also a contributing factor in explaining the early raveling observed in this section. The variations in WFVs were possible due to the inconsistent aggregate gradations and differing total AV contents at the surface of the pavement.

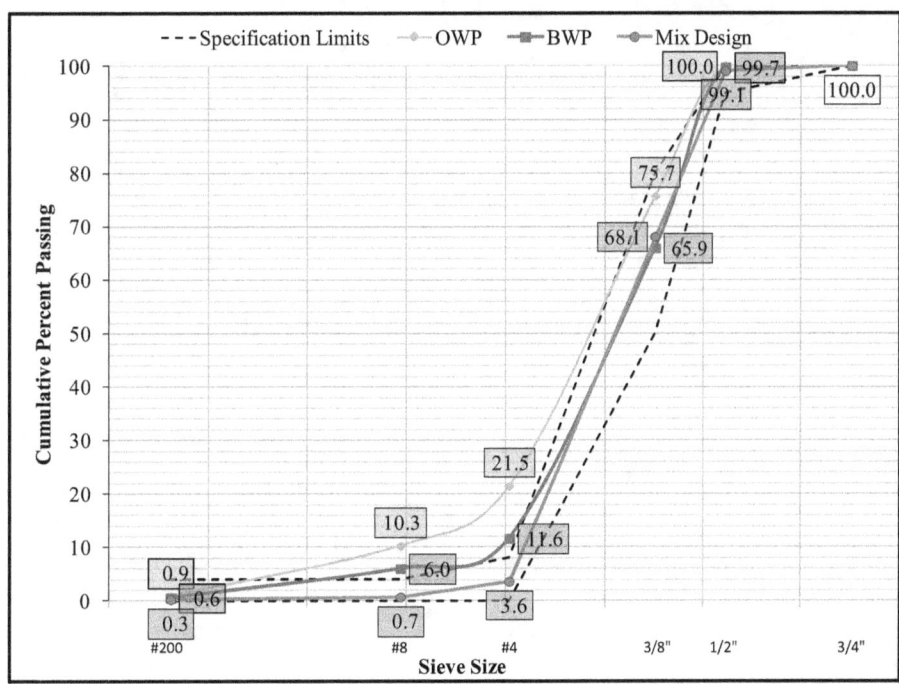
Figure 75. Aggregate Gradation after Ignition Oven for SL 289 FR in Lubbock.

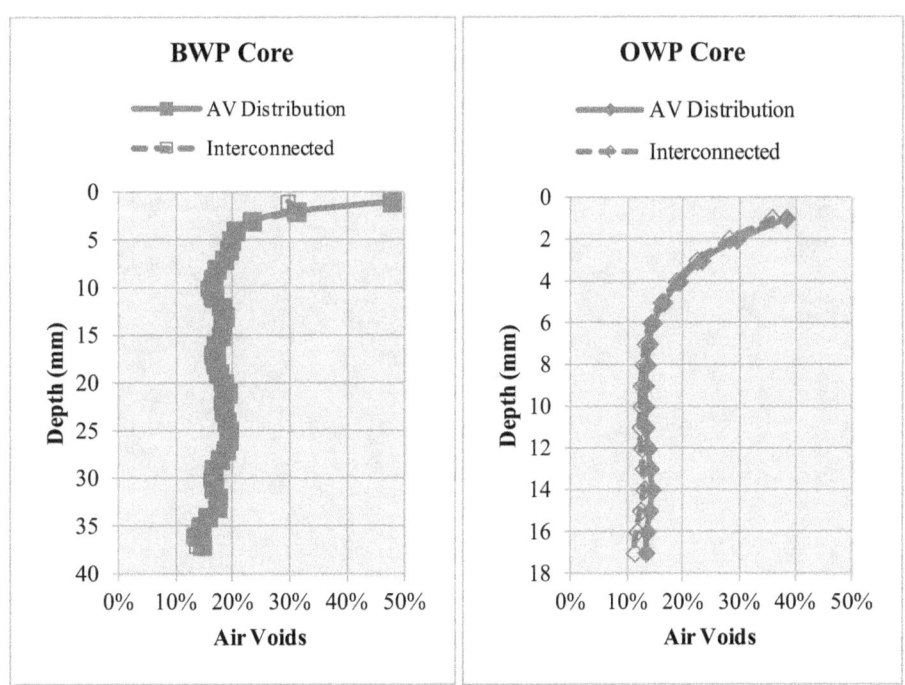
Figure 76. Total and Interconnected AV Distribution for Cores Obtained from SL 289.

CHAPTER 5:
COST-EFFECTIVENESS ANALYSIS

A conventional cost-benefit analysis (CBA) can be successfully applied to a highway project when a database is available for the cost of materials, construction, and performance of the roads being evaluated. Quantification of costs and benefits could be subjective for projects where safety and environmental elements are key components. For these types of projects, the cost-effectiveness analysis (CEA) has proven beneficial since it does not assign a monetary value to the attributes considered. By means of CEA, one can compare different alternatives without having to quantify their effects in dollars and incorporating local knowledge and expertise. This method groups the costs and benefits of the alternatives into categories. Several attributes characterize the performance of the alternatives within each category. Relative weights are then assigned to the attributes in each category and are also given to each category. The weighted sum of relative scores is computed and calibrated with available data and observations in an iterative process. The iteration continues until the results match the actual observations closely.

An alternative methodology that combines the advantages of both approaches is the generalized benefit/cost analysis (GBC). In this method, the attributes within each category are classified either as a generalized cost or as a generalized benefit. Generalized cost and benefit do not necessary have to be expressed in monetary terms. Then, the ratio of generalized benefits over generalized costs is obtained and the relative weights are established by minimizing the differences between the B/C calculations and the observations.

During the performance of this project, the framework for a CEA was developed because it matched the original technical objectives of the research project. Fifteen attributes grouped into four categories were identified. However, complete data from only 17 sections were available; therefore, the data were not considered sufficient for the calibration of a robust and reliable methodology. For that reason, the researchers developed a second, more simplified approach based on the GBC. The framework for the GBC analysis and the results are presented in the last section of this chapter.

COST-EFFECTIVENESS ANALYSIS APPROACH

In this section, the development of the CEA is described, step-by-step, and the selected categories and attributes are presented. It is important to note that this framework is presented only to document the work performed. However, the framework has not been calibrated due to the insufficient data available at this time.

Step 1: Categorizing Costs and Benefits into Categories

Costs and benefits of using Permeable Friction Courses (PFC) mixtures are categorized into four major groups:

1. Durability.
2. Functionality.
3. Safety.
4. Additional Costs and Benefits.

The overall performance of PFC mixtures is described in terms of *Durability and Functionality*. Durability describes the integrity of the road surface to maintain its structural integrity over time. Functionality describes longevity of specific functions that PFCs demonstrate over time. For instance, clogging of AV over time affects the functional life of PFCs. These two groups were initially assumed to have the half weight of the entire analysis, with a higher emphasis on functionality: Durability 20 percent and Functionality 30 percent.

The other half weight of the analysis is divided between *Safety* and *Additional Costs and Benefits*, with a higher weight for Safety. Saving lives on roadways is a high priority within all road agencies in the United States, particularly for TxDOT.

Step 2: Determining Attributes within Each Category and Assigning Relative Weights to Each Attribute

The second step in establishing the CEA framework for PFCs is determining attributes within each category and assigning relative weights to each attribute. Table 12 shows the attributes and preliminary weights assumed for the purpose of this analysis. Three to five attributes were selected in each category with even weights. These are initial values that need to be calibrated with input from TxDOT personnel.

Table 12. Categories and Attributes Used for the Cost-Effectiveness Framework.

Category	Attributes	
Durability (20%)	Raveling/Moisture Damage	33 %
	Bleeding/Fusion	33 %
	Performance Duration	33 %
Functionality (30%)	Noise (OBSI)	20 %
	Permeability (WFV)	20 %
	Texture (MPD w/ CTM) -(mm)	20 %
	Friction Number (w/ DFT)	20 %
	Skid Number (0-100)	20 %
Safety (30%)	Number of Wet Weather Accidents	33 %
	Number of Total Accidents	33 %
	Number of Wet Weather Days	33 %
Additional Costs and Benefits (20%)	Binder Type	25 %
	Binder Content	25 %
	Water Quality	25 %
	Relative Cost	25 %

Attributes in the Durability Category

Because of the high AV ratio, raveling and moisture susceptibility are reported as the most frequent cause of failure in PFC mixtures in Texas. This is due to aging of the binder and also to softening of the binder because of infiltration of fuel and oil drops into the porous structure of PFCs. In addition, inadequate asphalt content, aggregate gradation or insufficient compaction can accelerate raveling in PFCs. Conversely, PFC mixtures have the potential to eliminate or delay bleeding due to the porous structure. However, this may only be a temporary solution and may lead to clogging.

Performance duration is determined as the operational life of the road before the first major rehabilitation. To date, this attribute could be observed only for some sections but as time goes on, all the sections being monitored will have to be rehabilitated. For this reason, it is recommended that the monitoring of the PFC sections identified in this project continue.

Attributes in the Functionality Category

Functionality of PFC mixes is evaluated through five attributes: noise, permeability, texture, friction, and skid number. Although these attributes can be correlated, they were evaluated separately to achieve a more comprehensive analysis of functional performance.

Noise reduction and high drainage capacity are the two main functional characteristics of PFCs in Texas. PFCs are believed to reduce the tire/pavement noise level by 3 dBA as compared to dense-graded HMAs. Two key aspects to maintain a quiet PFC mix over the functional life of the project are the high porosity and surface texture (*38*). Although high porosity of the layer can be found using a permeability test, noise reduction capacity should be evaluated using a direct sound test. A permeability test can also be an indirect method of measuring splash and spray during wet weather conditions.

The texture of PFC mixtures is important both in terms of acoustical performance of the road surface and in tire/pavement friction. A relatively flat surface with voids in it (negative texture) has been proven to give better acoustical performance than a surface with protrusions above the surface (positive texture). Macrotexture and microtexture of the road surface play a key role in frictional properties and skid resistance of the road. While abrasion of the microtexture results in reduction in the low-speed skid resistance of the surface, reduced macrotexture causes reduction in the high-speed skid resistance of the surface.

Although some European studies have expressed serious concerns regarding skid potential and reduced frictional resistance of freshly paved PFCs in dry conditions, local studies reported almost the same friction numbers for PFCs and HMAs (see Table 13). These concerns can be related to the relatively thick binder film that exists on the surfaces of the aggregates of new PFCs. However, the traffic will abrade this binder film from the surface of the aggregate and polish the microtexture of the surface. Continuation of this process along with clogging of the voids and flushing of the surface may affect the macrotexture of PFC mixes and reduce the high-speed skid resistance of the road.

Table 13. Friction Measurements by McDaniel and Thornton 2005 (*39*).

Mix	Average Dynamic Friction Tester (DFT) Number (standard deviation in parenthesis)			International Friction Index (F_{60})
	20 kph	40 kph	60 kph	
PFC	0.51 (0.03)	0.45 (0.03)	0.42 (0.03)	0.36
SMA	0.37 (0.01)	0.31 (0.01)	0.29 (0.01)	0.28
HMA	0.52 (0.01)	0.47 (0.01)	0.44 (0.01)	0.19

Attributes in the Safety Category

Evaluation of the number of accidents represents an appropriate indicator for comparing the level of safety on roads with almost the same amount of traffic. One of the main purposes of the use of PFC layers is reduction in the number of wet weather accidents. However, due to the concerns pointed out regarding the change in the macrotexture and microtexture of PFC mixes, leading to potential reduction in skid resistance on the road, a potential increase in the rate of accidents may be expected in both wet weather and dry weather conditions. Thus, both the number of wet weather accidents and the total number of accidents were used for the safety attributes. Moreover, it is not reasonable to compare the number of wet weather accidents without taking into account the number of wet days. The higher number of wet days increases the risk of accidents happening on wet surfaces. Therefore, the average number of wet days per year is selected as the third attribute in safety group.

Attributes in the Additional Costs and Benefits Category

Initial cost is not the primary reason for choosing PFC mixes or not. According to a national survey conducted in TxDOT Project 0-4834, the cost makes up less than 1 percent of the reasons to discontinue the use PFCs. As a result, the actual cost of PFCs was not considered as an independent category, but the relative cost of PFC roads as compared to the conventional dense-graded mixtures was used. The maintenance cost for each road is taken into account in the Relative Cost attribute. In addition, some other factors in the Additional Costs and Benefits group may locally affect the decision of using PFCs such as Binder Type, Binder Content, and Drained Water Quality.

According to Item 342, both PG and AR binders can be used in PFCs. This item identifies the asphalt content for PG mixtures between 6 and 7 percent, and asphalt content for AR mixtures between 8 and 10 percent.

The last attribute considered is the impact of PFC on the quality of collected water off the road surface. Several researchers have shown the ability of PFCs to improve the quality of water runoff during rain events. The use of PFCs reduces the amount of splash and spray on the road, and consequently, reduces the contaminant washed off from the vehicle. In addition, the porous structure of PFCs can retain a portion of the generated pollutants. This structure may also perform as a filter to prevent transfer of the pollutant outside of the pavement.

Step 3: Defining Metrics and Scores for Each Attribute

The next step is to assign metrics and scores to each attribute. These metrics help to differentiate between the performance of each pavement section. For most attributes, recommendations and measures can be found in the literature regarding good or poor performance. However, since the purpose of CEA is to compare the pavement sections relative to one another, it was reasonable to define metrics based on the specific values obtained in this research project. For example, if a section has the minimum level of performance for some attribute, it receives the minimum score for that attribute. If another section has the maximum level of performance for the same attribute, it receives the highest score. The scores assigned to the metrics are defined on the scale of 0 to 1 (0 for the worst and 1 for the best).

Step 4: Use Weighted Sum to Determine Final Score for Each Project

Once weights are established for each attribute in the CEA, the final score for each project can be found by using the weighted sum formulation:

$$Project\ Score = \sum_{all\ groups} Category\ Weight \times Category\ Score$$

Since this project has four categories—Durability, Functionality, Safety, and Additional Costs and Benefits with relative weights of 20 percent, 30 percent, 30 percent, and 20 percent, respectively—the final score of each project will be defined using this formulation:

$$Project\ Score = 20\% \times Durabity\ Score + 30\% \times Functionality\ Score \\ + 30\% \times Safety\ Score + 20\% \times Additional\ C.\&B.\ Score$$

To find the score for each category, the weighted sum formulation should be used for the attributes:

$$Group\ Score = \sum_{all\ atributes} Attribute\ Weight \times Attribute\ Score$$

Once again, it is important to emphasize that the category and attribute weights presented in this chapter have not been calibrated due to the lack of sufficient data for the number of categories and attributes selected. For this reason, a generalized benefit/cost framework was developed, which use fewer variables.

DESCRIPTION OF THE RESEARCH PROJECT

For a better understanding of the selection of metrics and scores for each of the attribute, a brief description of the project is presented in this section. The purpose of this research project was to develop a database for the performance of PFCs. As a part of this project, researchers have monitored the field performance of 22 sections including a number of previous TxDOT projects along with new constructed projects. The monitoring program involves a comprehensive set of non-destructive field tests including:

- OBSI measurement.
- WFV test.
- GPR test.
- TxDOT Skid Trailer Test.
- CTM test.
- DFT.

Visual conditions surveys are performed on a regular basis to observe the overall performance of the sections. Table 14 shows the factors and levels considered in selecting the field sections. This table was the basis for selecting the PFC test sections in former TxDOT Project 0-5262. This helped the authors to integrate the already available database for the performance of PFC roads with the new data collected from the same sections and the newly constructed sections.

Table 14. Proposed Factors and Levels to Consider in Field Section Identification.

Factor	Levels
Binder/Modifier Type	PG76-XX, AR
Aggregate Type	Limestone, Sandstone, Granite, Gravel
Aggregate Gradation	CONSTANT w/Binder/Modifier Type = PG76-XX, AR, Fine Gradation
Aggregate Blending	100%, Class A, Blend, 100% Class B
Mixture Type	CONSTANT = PFC
Bonding Type	None, Tack Coat, Seal Coat, Thin-Bonded (Item 3000)
Texas Environmental Zone	Wet Warm (WW), Dry Cold (DC), Wet Cold (WC), Dry Warm (DW), Moderate (M)
Monitoring Date	CONSTANT = @ Construction & Spring/Summer 2010, 2011, & 2012 for New Field Sections CONSTANT = Spring/Summer 2009, 2010, 2011, & 2012 for Previous Field Sections
Field Evaluation	Field Water Flow; GPR; Noise – OBSI, roadside, impedance tube absorption test; TxDOT skid trailer, CTM, DFT
Coring Date	CONSTANT = As Needed (Expect Four Sections per Year in 2009, 2010, 2011, and 2012)
Coring Location	CONSTANT = Where Problem Identified

To take into account the effect of environmental conditions, the research group selected the test sections in five Texas climatic regions (see Figure 3). Table 15 presents the selected pavement sections. The 22 sections shown in this table fall into three main groups:

- Five non-PFC sections (one UTBHMWC and four dense-graded HMA): 8, 10, 13, 15, and 18.
- Five AR-PFC sections: 2, 3, 11, 15, and 21.
- 12 PG-PFC sections.

Table 15. Pavement Sections Selected for the Purpose of the CEA.

ID	Type	District	County	Route	Const. Yr.	Climate	Binder
1	PFC	Yoakum	Wharton	US 59	2007	WW	PG76-22
2	PFC	Houston	Brazoria	SH 288	2006	WW	AR
3	PFC	Austin	Bastrop	US 290	2007	M	AR
4	TBPFC	Paris	Hopkins	IH 30	2006	WC	PG76-22
5	PFC	Bryan	Robertson	SH 6	2009	WW	PG
6	PFC	Abilene	Taylor	IH 20	2006	DC	PG76-22
7	PFC	Abilene	Taylor	US 83	2006	DC	PG76-22
8	UTBHMWC	Wichita Falls	Wichita	SH 240	2008	DC	PG76-22
9	PFC	Waco	McLennan	SH 6	2005	M	PG76-22S
10	REF	Austin	Travis	SH 71	2008	M	PG76-22
11	PFC	San Antonio	Bexar	US 281	2005	DW	AR
12	PFC	Houston	Waller	SH 6	2005	WW	PG
13	REF	Yoakum	Wharton	US 59	2004	WW	PG70-22S
14	PFC	Corpus Christi	San Patricio	IH 37	2004	DW	PG76-22
15	PFC	Corpus Christi	Nueces	IH 37	2004	DW	AR
16	REF	Corpus Christi	San Patricio	US 77	2009	DW	PG
17	PFC	Waco	McLennan	IH 35	2003	M	PG76-22TR
18	REF	Paris	Hopkins	SH 154	2004	WC	PG64-22
19	PFC	Tyler	Van Zandt	IH 20	2007	WC	PG
20	PFC	Tyler	Smith	IH 20	2009	WC	PG 76-22TR
21	PFC	San Antonio	Bexar	US 281	2006	DW	AR
22	TBPFC	Houston	Fort Bend	SH 6	2004	WW	PG 76-22S

ATTRIBUTES METRICS AND SCORES

To exemplify the CEA approach, the research team ran a preliminary analysis with the weights given in Table 12. The next step was calibrating the method using the most recent field data and the opinion of TxDOT experts. For this preliminary analysis, researchers used the data collected in 2010.

Scoring the Noise Level Attribute

Figure 77 shows the results of the OBSI test on the field sections. The results are grouped by type of mixture. The first five sections from the left are the non-PFCs (reference sections), the middle bars are PG-PFC sections, and the bars in the right represent the AR-PFC sections. The rightmost bar in each group (in grey) represents the average value for each group. While the average value of sound intensity in PG-PFC sections is almost equal to the non-PFC sections, the average value for AR-PFC sections is lower. On average, AR-PFCs are quieter.

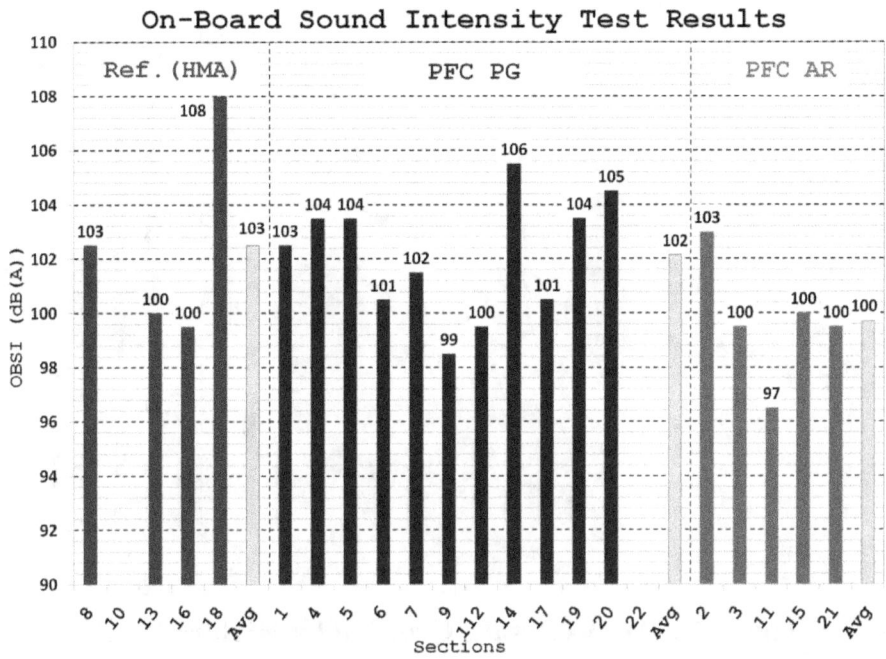

Figure 77. OBSI Noise Levels Grouped by Mixture Type.

These results were used to define the metrics for the noise attribute and scoring the sections. The noise level was divided into three ranges (see Table 16).

Table 16. Proposed Metrics for OBSI.

Metrics for The Noise Level Attribute		Score
<100 dB	Quiet (good)	1.00
100 to 103 dB	Average (fair)	0.50
>103 dB	Loud (poor)	0.00

NOTE: The metrics and scores presented in this document are recommendations of the research team and can be modified, if necessary.

Scoring the Permeability Attribute

Permeability of the field sections is evaluated via the WFV. Smaller numbers for WFV means faster drainage of water through the pavement surface and better permeability performance. Figure 3 illustrates the WFV numbers for the PG-PFC and AR-PFC sections. The rightmost bar in each group represents the average value for the sections in that group. PFC-PG sections show better performance than PFC AR sections.

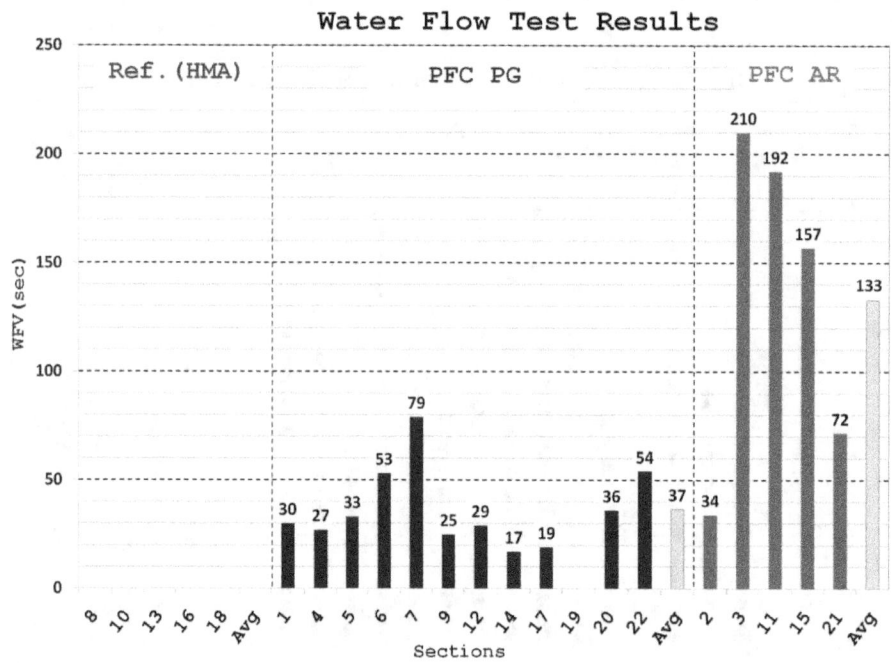

Figure 78. WFVs Results Grouped by Mixture Type.

According to the results presented in Figure 78, permeability performance of test sections is divided into the cases listed in Table 17.

Table 17. Proposed Metrics for WFV.

Metrics for The Permeability Attribute		Score
< 40 sec	Good	1.00
40 to 80 sec	Fair	0.50
> 80 sec	Poor	0.00

Since the HMA sections are almost impermeable, these are not able to drain off any water within reasonable time. Therefore, Figure 78 does not show WFV numbers for these sections. Correspondingly, the assigned score for these sections would be zero.

Scoring the Texture Attribute

The porous surface of PFCs does not allow the conventional sand-patch testing to be accurately performed. Therefore, the CTM test was used to determine the texture of the road surface in terms of MPD. Figure 4 shows the most recent CTM results. Unfortunately, the results were not collected for a number of sections; however, the available data show a significant difference between the MPD value of PFC sections and the MPD value of the reference sections.

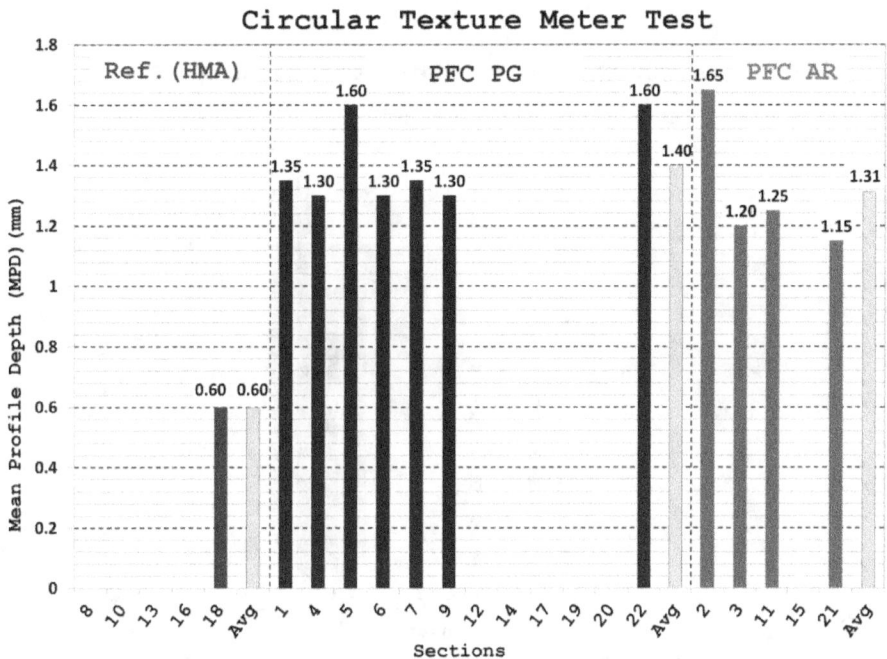

Figure 79. Texture Results Grouped by Mixture Type.

These results were used to determine the metrics and scores for texture. Table 18 lists the four groups to which the macrotexture of the roads was divided ranging from MPD < 0.5 mm (0.020 inch), the worst texture, to MPD > 1.5 mm (0.059 inch), the best texture. For those sections where CTM testing results were not available, the research team assumed the average value of the corresponding group for this preliminary analysis.

Table 18. Proposed Metrics for Texture.

Metrics for the Texture Attribute		Score
> 1.5	Good	1.00
1 to 1.5	Fair	0.67
0.5 to 1	Poor	0.33
< 0.5	V. poor	0.00

Scoring the Skid Resistance Attribute

Skid resistance was measured using two types of tests: TxDOT's Skid Trailer Test and the Dynamic Friction Tester (DFT). TxDOT's Skid Trailer test uses the procedure in ASTM E 274-06 and ASTM E 524-06. The coefficient of friction between the wheel and the pavement surface is calculated as the Skid Number (also called Friction Number, FN) on a scale of 0 to 100. A higher value for Skid Number means a higher coefficient of friction between the road and the tire. Figure 80 shows the results of the TxDOT Skid Trailer tests.

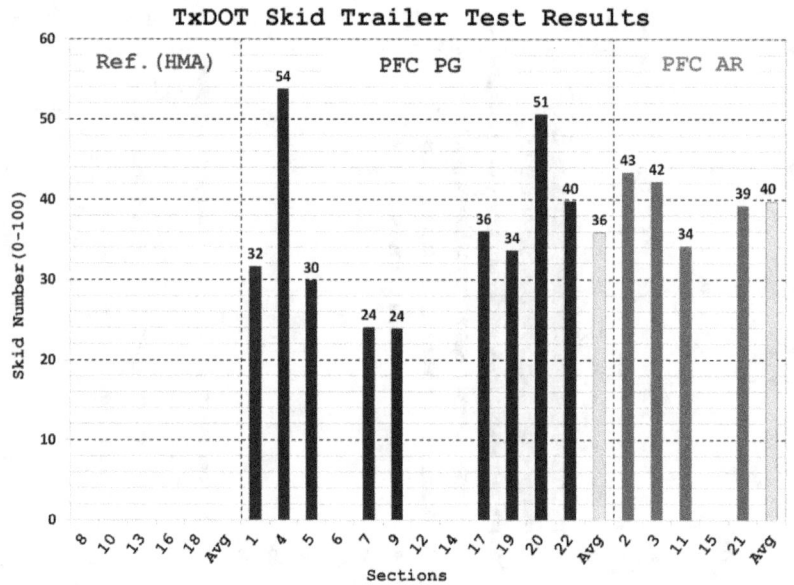

Figure 80. SN Results Grouped by Mixture Type.

The Skid Trailer Testing data were not available for several sections, so DFT measurements were used. DFT measures the friction performance of the road in terms of a number between 0 and 1. DFT is a portable test but it requires traffic control. Figure 81 shows the data collected with the DFT. By comparing Figure 80 and 81, one can observe that for most of the sections the results of TxDOT's Skid Trailer Test are consistent with the results of the DFT. Therefore, in the absence of direct test data for either one of these tests, the available information from the results of the other test can be used. Furthermore, analysis of the results shows that, on average, the skid resistance of the PFC-PG sections is similar to the PFC-AR sections, but lower than that of the HMA group. Table 19 presents the metrics for FN and SN used for the purpose of the CEA.

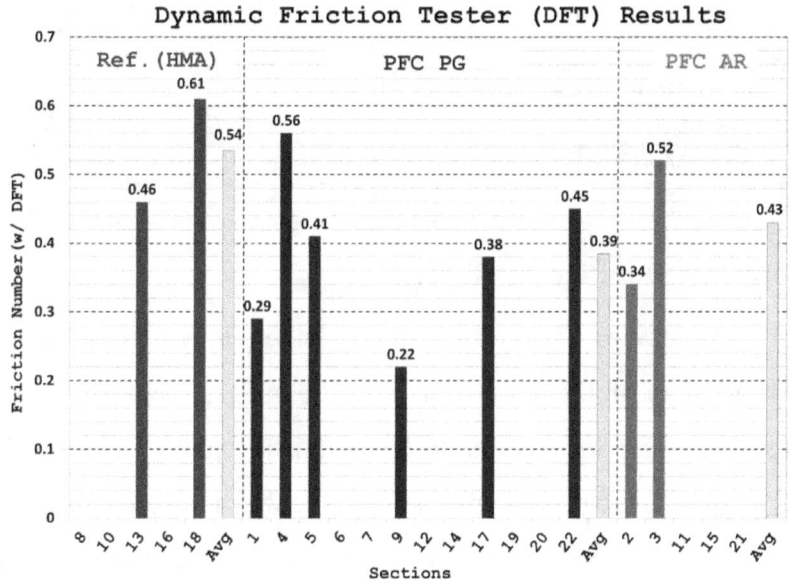

Figure 81. DFT Results Grouped by Mixture Type.

Table 19. Proposed Metrics for Friction Number and Skid Number.

Attribute	Score		
Friction Number (w/ DFT)	> 0.45	V. good	1.00
	0.35 to 0.45	Good	0.67
	0.25 to 0.35	Fair	0.33
	< 0.25	Poor	0.00
Skid Number (0–100)	> 40	V. good	1.00
	35 to 40	Good	0.75
	30 to 35	Fair	0.50
	25 to 30	Poor	0.25
	< 25	V. poor	0.00

Scoring the Raveling/Moisture Susceptibility and Bleeding/Fusion Attributes

TxDOT's PMIS was used to collect the data for these attributes. According to the PMIS Raters Manual FY 2010, raveling and bleeding distress data are recorded using a code between 0 (for no distress) and 3 (for higher than 50 percent area of distress). For the CEA, this scale was used for the metrics of the raveling and bleeding attributes. The lowest score, i.e., 0, is assigned for Code 3 and the highest score, i.e., 1, is assigned for Code 0 (see Table 20).

Table 20. Proposed Metrics for Raveling and Bleeding.

Percent Area of Distress		PMIS Rating Code	Score
0	None	0	1.00
1 to 10 %	Low	1	0.67
11 to 50 %	Medium	2	0.33
> 50%	High	3	0.00

Scoring the Performance Duration Attribute

The performance duration of each road section is determined as the operational life of the road before the first major rehabilitation. The service life of PFCs depends on the level of distresses and ranges between 6 to 10 years. In Texas, this is typically between 5 and 8 years. For the purpose of this project, the highest performance score was assigned to road sections with performance of longer than 10 years. The lowest score was assigned to road sections with less than 5 years of life. Then, the road sections are classified in four categories between 5 and 10 (see Table 21).

The oldest section in the database is Section 17 (PG-PFC), which was constructed in 2003, and the newest are Sections 5 (PG-PFC), 16 (HMA), and 20 (PG-PFC), which were constructed in 2009. For those sections that have not been rehabilitated, the performance duration is an underestimation of their performance. As the time progresses, these estimates will become more accurate.

Table 21. Proposed Metrics for Performance Duration.

Metrics for Performance Duration		Score
V. good	>10 years	1.00
Good	8 to 10 years	0.67
Fair	5 to 8 years	0.33
Poor	<5 years	0.00

Scoring the Safety Attribute

TxDOT's CRIS was used for collecting the data on crashes on the field sections. The collected data for all sections were compared with each other to determine the different quartiles. Sections in the first quartile of accident numbers (lower numbers) are assumed to perform well and assigned the highest score, i.e., 1. The next quartiles are scored accordingly (see Table 22).

Table 22. Proposed Metrics for Number of Accidents.

Metrics for the Number of Accidents		Score
< 1st quartile	Good	1.00
1st to 2nd quartile	Fair	0.67
2nd to 3rd quartile	Poor	0.33
> 3rd quartile	V. poor	0.00

The number of wet weather days per year can be determined using the nearest weather station to each field section. However, in the absence of accurate data from weather stations, the climatic regions can be used as a tentative substitute. It would be expected that the risk of wet weather accident is the highest in the WC climatic region and the lowest in the DW climatic region. Consequently, the highest score (i.e., 1) is assigned to the roads located in the WC region and the lowest score (i.e., 0) is assigned to the roads in the DW region. It is, however, recommended that more accurate data are collected for subsequent analysis.

Table 23. Proposed Metrics for the Number of Wet Days.

Metrics for the Number of Wet days	Score
WC	1.00
WW	0.75
M	0.50
DC	0.25
DW	0.00

Scoring the Additional Costs and Benefits Attribute

The following metrics and scores used in the Additional Costs and Benefits attribute are listed in Table 24:

- Reference (HMA) sections were assigned the highest score in the Binder Type attribute because of the lower initial cost.
- The limits used for the Binder Content attribute are 6 percent and 8 percent, which are the minimum optimum asphalt content for PG-PFC and AR-PFC, respectively. However, according to Item 342 of TxDOT Specifications, the asphalt content should be between 6 and 7 percent for PG-PFCs, and between 8 and 10 percent for AR-PFCs. Therefore, typically, the AR sections were assigned the lowest score, and HMA sections were assigned the highest score.

- The PG-PFC sections were assumed to have the highest benefit in terms of water quality, followed by the AR-PFCs and then the HMA. However, more research is needed to understand the difference between the qualities of water drained off the AR-PFCs and those drained off the HMA sections.
- The basis for the relative cost is the cost of a conventional dense-graded mixture. Therefore, the relative cost of the reference sections (HMA) is assumed the minimum. The additional cost incurred for PFCs can vary locally (due to locally available aggregates, local regulations and subsidies, the use of fibers, etc.). In some areas, PG-PFCs are considered cheaper options; in other areas, AR-PFCs are cheaper. Moreover, the PMIS database provides the yearly total maintenance cost spent on each road section. These data can also be used to compare the relative cost of the field sections.

Table 24. Proposed Metrics for Additional Costs and Benefits.

Metrics for the Additional Costs and Benefits Group			Score
Binder Type	Neat (no PFC)		1.00
	AR		0.50
	PG		0.00
Binder Content	< 6 %	Good	1.00
	6 to 8 %	Fair	0.50
	> 8 %	Poor	0.00
Water Quality	Good (PFC PG)		1.00
	Fair (PFC AR)		0.50
	Poor (HMA)		0.00
Relative Cost	Relative Cost < 1	Good	1.00
	1 < Relative Cost < 1.1	Fair	0.67
	1.1 < Relative Cost < 1.5	Poor	0.33
	Relative Cost > 1.5	V. poor	0.00

GENERALIZED BENEFIT/COST ANALYSIS

Due to insufficient data for the number of attributes identified, the researchers decided to apply an economic analysis method based on generalized benefits and costs. The basic framework that they followed consisted of the following steps:

- Select only relevant and available attributes based on 17 sections (Table 25).
- Categorize the available attributes into Generalized Benefits (Xi) and Costs (Yi).

- Establish metrics for the attributes. In most cases, the research team used the same metrics as the ones described in the previous section of this chapter. Those that were modified are summarized below.
- Rate the 17 sections from the best performing projects to the worst performing project. The members of the research project team that were involved in the field evaluation ranked the 17 available PFC projects (see Table 25).
- Estimate the parameters to minimize the differences between the calculated generalized benefit cost ratio and the observations.
- Adjust the estimated parameters based on a pre-established threshold that differentiates between acceptable and non-acceptable performance.

Table 25. Final List of Pavement Sections Used for the B/C Analysis.

0-5836 ID	District	Route	Binder	SAC	Rank
112	Houston	SH 6	PG TR	A	1
17	Waco	IH 35	PG TR	A	2
20	Tyler	IH 20	PG TR	A	2
4	Paris	IH 30	PG	A	3
5	Bryan	SH 6	PG	A	3
19	Tyler	IH 20	PG	A	3
1	Yoakum	US 59	PG	B	4
6	Abilene	IH 20	PG	B	5
11	San Antonio	US 281	AR	A	5
15	Corpus Christi	IH 37	AR	B	5
21	San Antonio	US 281	AR	A + B	5
2	Houston	SH 288	AR	A + B	6
3	Austin	US 290	AR	A	6
7	Abilene	US 83	PG	B	7
25	Wichita Falls	LP 473	PG	A + B	7
9	Waco	SH 6	PG	B	8
14	Corpus Christi	IH 37	PG	A + B	8

The governing equation is the following:

$$B/C\ Ratio = \frac{\alpha_1 X_1 + \alpha_2 X_2 + \cdots + \alpha_n X_n}{\beta_1 Y_1 + \beta_2 Y_2 + \cdots + \beta_m Y_m}$$

Where:

X_i : Attributes in the Benefits Group.

Y_j : Attributes in the Costs Group.

α_i, β_j : Estimated parameters.

The following attributes were classified as benefits:

- Raveling.
- Performance
- Noise
- Drainability
- Skid number

There were classified as generalized costs:

- Number of wet weather accidents.
- Binder type.
- Binder content.
- Aggregate classification.

The analysis indicated that, for the available data, the following variables were not significant in explaining the observed performance and were removed from the model:

- Raveling.
- Noise.
- Number of wet weather accidents.
- Binder content.

The final model explained 87 percent of the observed performance based on only four variables:

- Performance.
- Drainability.
- Binder type.
- Aggregate classification.

The following equation gives the final model and it is presented against the data in Figure 82. The variables Performance, Drainability, Binder Type, and SAC in the equation were scored according to the values listed in Table 26.

$$B/C\ Ratio = \frac{2.42 Performace + 0.32 Drainability}{1 + 0.12 BinderType + 0.29 SAC}$$

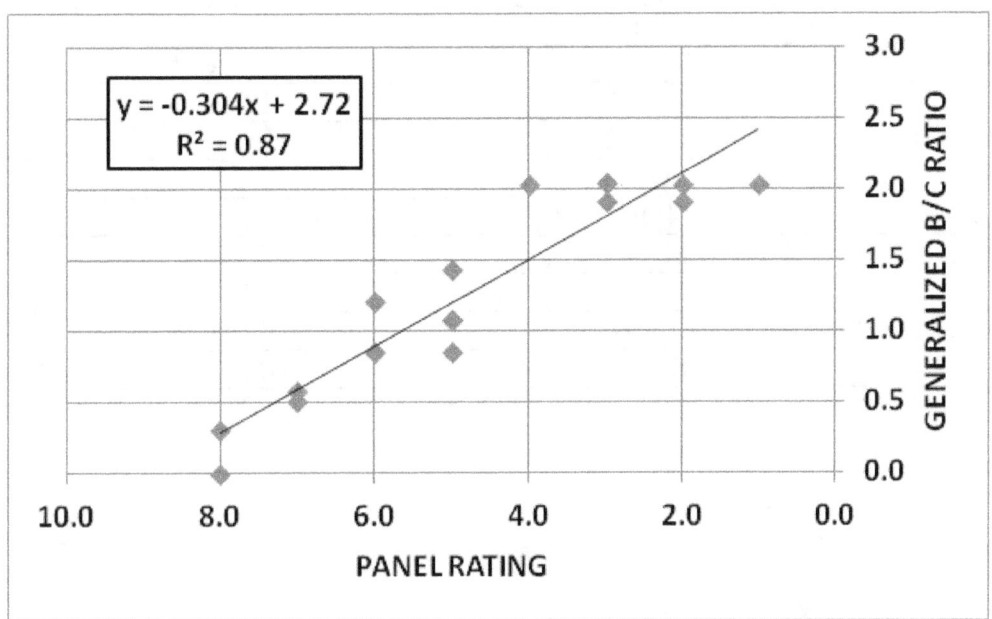

Figure 82. Correlation between B/C Ratio and Panel Rating.

Table 26. Scores for the Generalized Benefit/Cost Analysis Model Variables.

Performance	Rank	Score
Very Good	1, 2, 3	1.0
Good	4	0.7
Fair	5, 6	0.5
Poor	7	0.3
Very Poor	8	0.0
Drainability (WFV)	Quality	Score
< 40 sec	Good	1.0
40 to 80 sec	Fair	0.5
80 sec	Poor	0.0
	Type	Score
	AR	1.0
Binder	PG TR	0.5
	PG	0.4
	Neat	0.0
	SAC	Score
	A	1.0
Aggregate Classification	A + B	0.5
	B	0.0

CHAPTER 6:
GUIDELINES ON DESIGN, CONSTRUCTION, AND MAINTENANCE OF PERMEABLE FRICTION COURSES

Based on the results of this study, the guidelines developed in study 0-5262 have been updated and are presented in the following. PFCs are defined in Item 342 of the 2004 TxDOT Standard Specifications book as a surface course of a compacted permeable mixture of aggregate, asphalt binder, and additives mixed hot in a mixing plant.

BENEFITS OF PFC

The most important benefit of PFC is increased safety during wet weather. The open void structure aids in the drainage of water and preservation of the surface friction. Benefits of PFC include the following:

- Reduced risk of hydroplaning and wet skidding.
- Decreased splash and spray.
- Reduced tire/pavement noise.
- Improved ride quality.
- Improved visibility of pavement markings at night and in wet weather.
- Cleaner storm water runoff compared to dense graded mixes.

USES OF PFC

PFC can be used in new construction, major rehabilitation projects, and maintenance overlays. It is typically used as a wearing course over dense graded mixtures or Portland cement concrete pavements. PFC can minimize hydroplaning potential by providing drainage channels for water to flow beneath the pavement surface. This will also minimize splash and spray due to rain by reducing the surface water and glare at night during wet weather resulting in better visibility. While effective in removing water from the pavement surface, it is not a solution for correcting cross slopes.

PFC should not be used to correct severe rutting or depressions in the underlying pavement. These depressions will allow water to accumulate and may accelerate pavement damage. Ruts and depressions should be leveled with dense-graded mix prior to overlaying with PFC.

PFC can also mitigate flushing and bleeding problems. The open AV structure allows the absorption of the free asphalt, which may alleviate the bleeding pavement. This, however, may only provide a short-term solution and may not prevent rutting or completely address the source of the distress.

Some of the areas where PFC should not be used include the following:
- Crossovers and Driveway Turnouts – These areas generally require a significant amount of hand work when placing the mix, and PFC does not lend itself to that level of workability.
- Muddy and Sandy Areas – Any locations where mud or sand can be trafficked from unsurfaced driveways or side roads can clog the PFC.
- Localized Areas to be Removed and Replaced – PFC should not be placed in areas where a bathtub effect may be created. When sections of dense-graded pavement are removed, dense-graded mix should be used as a replacement material before overlaying with PFC.

Examples of some of the successful uses of PFC around the state include:
- Urban interstates as shown in Figure 83 where the left portion of the photo shows the reduced splash and spray associated with the PFC compared to the dense-graded mix on the right half of the photo.
- Rural interstates (Figure 84).
- Rural US highways to mitigate bleeding surface treatment (Figure 85).
- Accident-prone severe curves (Figure 86).
- US highways including areas with stop-and-go intersections as shown in the Lufkin District in Figure 87. This PFC is more than 8 years old and is still in excellent condition.

Figure 83. PFCs in Urban Areas to Reduce Splash and Spray (San Antonio IH 35).

Figure 84. PFCs on Rural Interstates (Odessa District IH 10, AR Underseal Followed by AR-PFC).

Figure 85. US 183 Brownwood District (PFC Used to Remediate Bleeding Surface Treatment).

Figure 86. Lufkin District US 59 Cloverleaf (PFC Used to Remediate Wet Weather Accidents).

Figure 87. Lufkin District US 59 through Diboll.

COST CONSIDERATIONS

PFC is more expensive per ton than conventional dense-graded mixtures, but the unit weight of the mix in-place is lower, which can offset this cost. Layer thickness ranges from between 1 and 2 inches. The thicker mat can accommodate more water during heavy rain events and may also serve to better attenuate traffic noise.

If the mat thickness is only 1 inch, consideration may be given to not paving the shoulders. It is common practice in several districts to extend the PFC beyond the main lanes only 2 feet into the shoulder, which can reduce cost.

Once a PFC surface has reached the end of its life, it must be removed from the pavement surface. Because it can hold water, it should never be overlayed or seal coated.

MATERIALS AND DESIGN

Aggregates

One of the primary reasons for using PFC is to improve wet-weather safety conditions. Aggregate properties such as texture, shape, size, and resistance to polish are all factors affecting skid resistance. Under the department's Wet Weather Accident Reduction Program (WWARP), aggregates are classified into four categories (A, B, C, or D) based on a combination of frictional and durability properties. Either Class A or B aggregates may be used for PFCs based on the friction demand assessment made by the engineer. When Class A is required, blending of A and B is allowed if at least 50 percent of the material retained on the No. 4 sieve comes from the Class A source. Proposed new specifications for 2013 will allow the option of requiring 100 percent Class A material.

This research study has shown that PFCs constructed of both Class A and Class B aggregates have performed very well. Initially there were concerns that the Class B aggregates would degrade under traffic loadings; however, this is not the case. Several of the Class B PFC projects monitored in this study maintained excellent drainage capacity in the wheelpaths with no indication of rutting or aggregate degradation. However, the Class B PFCs showed poorer skid resistance than those constructed with Class A. While a PFC serves to remove water from the pavement surface thereby mitigating the potential for hydroplaning, the skid resistance of the surface is highly dependent on the frictional properties of the aggregates.

Based on the type of asphalt selected, master aggregate gradation bands are provided in Item 342. Aggregates must also meet requirements in Item 342 that include coarse aggregate angularity, deleterious materials, soundness, abrasion resistance, and flat and elongated particles. The gradation band for AR mixtures generally produces a finer mix than that for PG mixtures. Some of the AR projects monitored in this study did not maintain the drainage capability expected of PFC. This may be related to the aggregate gradation specified for AR mixtures, which is not as open as the PG gradation. In addition, the AR mixtures have 8.0 to 10.0 percent binder that can fill the air voids in the mix as opposed to the PG mixtures, which have 5.5 to 7.0. Proposed new specifications for 2013 will incorporate a coarser gradation for AR mixtures that may improve the drainage capability of these mixes.

The finer aggregate gradation used for AR mixtures may also be contributing to quieter noise measurements observed on some of these pavement surfaces.

Asphalt Binder

The following two types of binders are allowed in the specification:
- A Type I or II asphalt rubber (AR) defined in Item 300.2.I with a minimum of 15 percent by weight of asphalt of Grade C or Grade B crumb rubber defined in Item 300.2.G.
- A PG asphalt with a minimum high temperature grade of PG 76-XX defined in Item 300.2.J with a minimum of 1.0 percent of lime by weight of dry aggregate and a minimum of 0.2 percent of cellulose or mineral fibers by weight of mixture.

The AR mixtures evaluated in this study in general did not exhibit drainage properties as good as the PG mixtures. Item 342 allows the use of either a Grade C or Grade B crumb rubber, and most districts have used the Grade B. The Odessa District, however, uses a Grade C (which is finer than Grade B) and generally reports that their PFCs maintain drainage capability. This may indicate that the finer grind of rubber reduces the contribution of the rubber particle to closing up of the voids.

There are generally two types of PG binders that have been used in PFCs in Texas: PG 76-22 and PG 76-22TR. Excellent performance has been observed with both of these binders.

Mix Design

To obtain the benefits described above, a mix design system that produces both a functional and durable PFC mixture is required. In Texas, the PFC mix design is currently defined in TxDOT Test Method Tex-204-F, Part V, and material requirements are defined in Item 342 of the 2004 TxDOT Standard Specifications book (*1*).

Following selection of materials, two replicate specimens [150 mm (6 inch) in diameter by 114 mm (4.5 inch) in height] at three selected asphalt contents are mixed, oven-cured for 2 hours at the compaction temperature, and compacted in a Superpave Gyratory Compactor (SGC) at an N_{design} of 50. The three asphalt contents differ by 0.5 percent, and the minimum optimum asphalt content for PG and AR asphalts, respectively, must be 6.0 percent and 8.0 percent. According to Item 342, the asphalt content must be between 6 and 7 percent for PG mixtures and between 8 and 10 percent for AR mixtures. An optimum asphalt content is then selected based on the target laboratory density specified (between a suggested limit of 78 and a maximum of 82 percent according to Item 342 or equivalently between total AV contents of 18 and 22 percent evaluated using dimensional analysis) and the minimum requirements provided.

Next, specimens at the selected optimum binder content are produced for an evaluation of draindown (Tex-235-F), moisture susceptibility (Tex-530-C), and durability (Tex-245-F). The optimum mixture must have a maximum draindown of 0.2 percent, where draindown is defined as the ratio of: the change in the weight of paper plate that the mixture is allowed to drain onto from a wire mesh basket at the plant mixing temperature for 1 hour to the original specimen weight. The moisture susceptibility of the optimum mixture is determined by boiling the loose mixture in water for 10 minutes and visually evaluating the percentage of stripping after 24 hours. The percentage of stripping after 24 hours is reported for comparison during production, and no requirement is provided in the TxDOT specification. The engineer may reduce or waive the sampling and test requirements for the boil test based on a satisfactory test history.

Finally, the durability of the optimum mixture is evaluated based on the percentage of Cantabro loss, where Cantabro loss is defined as the change in weight of the specimen before and after an abrasion test divided by the original specimen weight. The test involves placing a compacted specimen into the Los Angeles abrasion equipment without the steel balls and rotating the apparatus for 300 revolutions at 30 to 33 revolutions per minute. Item 342 suggests a maximum Cantabro loss value of 20 percent, but this value is reported for information only.

Item 342 of the 2004 TxDOT standard specifications integrates aging of the binder, but only during production. Aging ratio is defined as the ratio of the high PG temperature Dynamic Shear Rheometer (DSR) parameter ($G^*/\sin \delta$) of the extracted and recovered binder sample, and this same parameter evaluated on the original unaged binder. A maximum aging ratio value of 3.5 is specified.

The construction of PFCs, in general, utilizes the current techniques applied to the construction of dense-graded mixes. However, some special considerations should be implemented throughout the process.

CONSTRUCTION

Mixture Production

Moisture Considerations

As in the production of dense-graded mixtures, PFC production requires special attention to aggregate moisture control. The mixing time and temperature should be controlled so that substantially all of the moisture is removed from the mix before discharging from the plant. TxDOT specifications (Item 342) from 2004 require that the mixture contain no more than 0.2 percent moisture by weight, which should ensure better control of mixing temperature and a more homogeneous mixture.

Addition of Fibers

When asphalt rubber binders are specified, fibers are not incorporated. Cellulose or mineral fibers should be used when a PG 76-XX binder is specified. Conventional asphalt plants can be adapted to allow the incorporation of fibers with the installation of a fiber feed device as shown in Figure 88.

In batch plants, bags of fiber can be added directly into the pugmill where the bags melt, and the fiber is distributed into the mixture. When using a batch plant to produce mixtures with mineral fibers or cellulose fibers, it may be necessary to lengthen both the dry and wet mixing times to ensure fiber distribution. Drying time may also need to be increased if the production temperature is lower than that used for other types of mixtures.

Mixing Temperatures

The binder supplier should be consulted to determine the appropriate plant mixing temperature. Minimum mixing temperatures must be maintained to ensure that the mix reaches the roadway at a temperature that provides for ease of placement. In addition, the maximum mixing temperature must also be monitored to prevent draindown of the binder. TxDOT specifications require that the maximum temperature not exceed 178°C (350°F) prior to shipping the mix from the plant and that the mixture shall not be placed at a temperature below 138°C (280°F).

TTI researchers are not aware of any research or construction projects employing the use of warm mix asphalt (WMA) technologies for PFC. While WMA could improve the ease of placement and workability of PFC mixtures, it may have a negative impact on binder draindown.

Figure 88. Introduction of Fibers into Drum on IH-30 Project in Paris District.

Mixture Storage and Transportation

Because PFCs may be prone to draindown, some DOTs limit mixture storage and transportation times. TxDOT requires that PFC mixtures not be stored for a period long enough to affect the quality of the mixture, nor in any case longer than 12 hours. Thus far, draindown of the binder has not been reported as a problem in the construction of PFCs in Texas.

Tarps are necessary to avoid crusting of PFCs during transportation. Insulated truck beds for transportation of PFC are required by some agencies. Item 342 does not require the use of tarps and insulated truck beds, although several TxDOT districts recommend that they be required on the job. Truck beds should be prepared for transportation by using a full application of an asphalt release agent.

Surface Preparation

PFCs should not be considered as a layer to correct profile distresses or any kind of structural distress. Before placement of PFC, the pavement surface should be corrected to avoid zones that allow water accumulations (e.g., zones with permanent deformation). Lateral and longitudinal drainage of the underlying surface must be provided to guarantee adequate water discharge from the PFC.

Many TxDOT districts prefer the surface directly beneath the PFC to be a seal coat. This ensures an impermeable membrane to protect the underlying layers from surface water intrusion. The seal coat also helps to provide a good bond between the PFC and underlying surface, particularly for Portland cement concrete pavement surfaces. Some districts recommend the use of an asphalt-rubber underseal when placing PFC on Portland cement concrete to ensure a good bond. PFCs have been placed over all types of pavement surfaces in Texas: including dense-graded mixes, SMA, crack attenuating mixes (CAM), and seal coated surfaces.

It is important to have an adequate tack coat to bond the PFC to the underlying surface. A good tack coat can also help to seal the surface from the intrusion of water from the surface. Item 342 states that the tack should be uniform and should be applied at a rate between 0.04 and 0.10 gal/sy residual asphalt as directed by the engineer. If there is a new seal coat underneath the PFC, some districts do not require a tack, although the Odessa District still required a tack coat on the new seal shown in Figure 89 prior to placement of the PFC.

Figure 89. Application of PFC on Tacked, Newly Seal Coated Surface of IH 10 in Odessa District.

Figure 90 shows the placement of PFC on IH 30 in the Paris District. In this operation, the laydown machine is equipped with an emulsion tank that is sprayed immediately in front of the hot mix. This allows for a heavier and perhaps more uniform application of tack since construction vehicles do not drive on the tacked surface. This type of PFC is known as a thin-bonded PFC and is governed by Special Specification.

Figure 90. Laydown Machine with Tank for Applying Tack on IH-30 in the Paris District.

Any edge clearing (Figure 91) should be performed prior to placement of the PFC. These fines on the outside edge of the mat can cause the mixture to clog near the edge.

Figure 91. Edge Clearing that Can Cause Clogging of PFC Restricting Lateral Water Flow.

Mixture Placement

Material Delivery

It is very important to monitor the temperature of PFC as delivered to the roadway. Any cold spots will form lumps in the mix and must be removed. Some districts recommend the use of a material transfer vehicle (MTV) to minimize the need to remove large chunks of mix. Even with the MTV, there may still be small chunks of mix requiring removal and patching from the mat as shown in Figure 92, which is in the Houston District on a project paved in early fall.

Figure 92. MTV Paving Operation (left photo) and Areas in the Mat Where Cold Chunks Required Removal and Patching (right photo) on US 288 in the Houston District.

The windrow pickup process tends to exhibit more thermal segregation for PFC. Several districts report that while this process can be used on hot, summer days, it should not be used for PFC mixes on cooler days. A windrow pickup process used in the Yoakum District is shown in Figure 93. Even on an August day in Yoakum, the end of the windrow portion representing the end of the truckload sometimes formed large chunks, which required removal with a front end loader.

Figure 93. Windrow Pick-Up Process on US 59 in the Yoakum District with Cold Lumps Forming at the End of the Windrow.

Item 342 requires that the mix delivered to the paver not drop below 138°C (280°F) and thermal profiles are required for each sublot. The Austin District personnel report that they prefer the mix to be at 164°C (325°F) as it is coming out of the trucks.

Acceptable paving conditions in the United States are commonly defined as a minimum air temperature of 16°C (60°F). Although this limit is used by most agencies, there are some exceptions. Florida, for example, requires a minimum air temperature of 7°C (45°F). TxDOT requires a minimum roadway temperature of 21°C (70°F).

Paver Operations

To produce a smooth surface, the paver should advance continuously with minimal stoppages. Districts report that this type of mix will cause a bump in the mat at each location where the paver stops. Any bumps and surface depressions left in the PFC mat are more difficult to correct than in dense graded mixes. When asphalt pavers with extendible screeds are used, researchers recommend auger extensions to avoid irregular distribution of mixture between the center and the edge of the paver. The use of a vibratory, hot screed is needed to avoid pulling excessively on the material and diminish the necessity of raking, which can cause areas with lower voids or, more likely, uneven void distribution across the pavement. In addition, raking can generate unsightly surface texture and poor aesthetics, which cannot be rolled out with compaction.

Initiation of mixture placement is recommended on the low side of the paving area to avoid accumulation of water (from the rollers or surface water) onto areas to be paved. It is desirable to minimize or even avoid mixture handworking.

If the PFC mat is to be placed in the main lanes only, thicker mats may require a taper to join the grade of the existing shoulder. If tapering is required, the Beaumont District recommends the use of a special-type milling machine to mill in the tapers due to the workability difficulties in constructing a taper with the paver for this type of mix. The Yoakum District was able to construct the taper for the notched wedge joint as shown in Figure 94. The smaller roller attached to the paver to roll the taper required a worker to constantly apply a release agent to minimize mixture pickup.

Figure 94. Notched Wedge Joint Construction in PFC.

Handwork on PFC mixtures is difficult to impossible. Experienced districts recommend staying away from crossovers and bridge ends when paving with PFC mixtures and instead paving these areas with dense-graded mix.

Special attention to placement and compaction temperatures is required since this mixture is constructed using modified binders and may be placed in a thinner mat than dense-graded mixes. These thin layers cool faster and allow less time for compaction. Some agencies specify a nominal thickness of 2 inches to maximize sound attenuation, spray reduction life, water storage capacity, and compaction time.

Material Compaction and Joint Construction

Static steel-wheel rollers are required for the compaction of PFCs. Pneumatic rollers must not be used since their kneading action reduces the mixture drainage capacity by closing surface pores. Minimal compaction is required to seat the mixture without excessive breakage of the aggregate and to provide a smooth surface and uniform texture. Roller drums should be thoroughly moistened with a soap-and-water solution to prevent adhesion. Only water or an approved release agent may be used on rollers, tamps, and other compaction equipment.

Typically, two to four passes (within the adequate range of temperature) with an 8- to 9-ton tandem roller are adequate to complete the compaction process on thin layers. There should be a balance between achieving the needed compaction to ensure durability of the mixture without degrading the aggregate as well as providing a minimal amount of compaction to achieve drainability.

TTI researchers recommend the use of Tex-246-F (Figure 14) to verify that the compacted mixture has adequate permeability. The test evaluates the time required to discharge a given volume of water channeled onto the pavement surface through a 150 mm (6 inch) diameter opening. This time corresponds to the WFV expressed in seconds.

Figure 95 shows how the field water flow values change with each roller pass and how two different mixes (both constructed under Item 342) can behave very differently. To ensure adequate permeability, the field water flow value should not exceed 20 seconds (*28*). For the US 290 mixture shown in Figure 95, this water flow value of 20 seconds corresponds to a compaction effort of not more than four passes. For the US 59 mix in Yoakum shown also in

Figure 95, the field water flow value did not change significantly after the first two passes. Mixture differences are shown in Table 27.

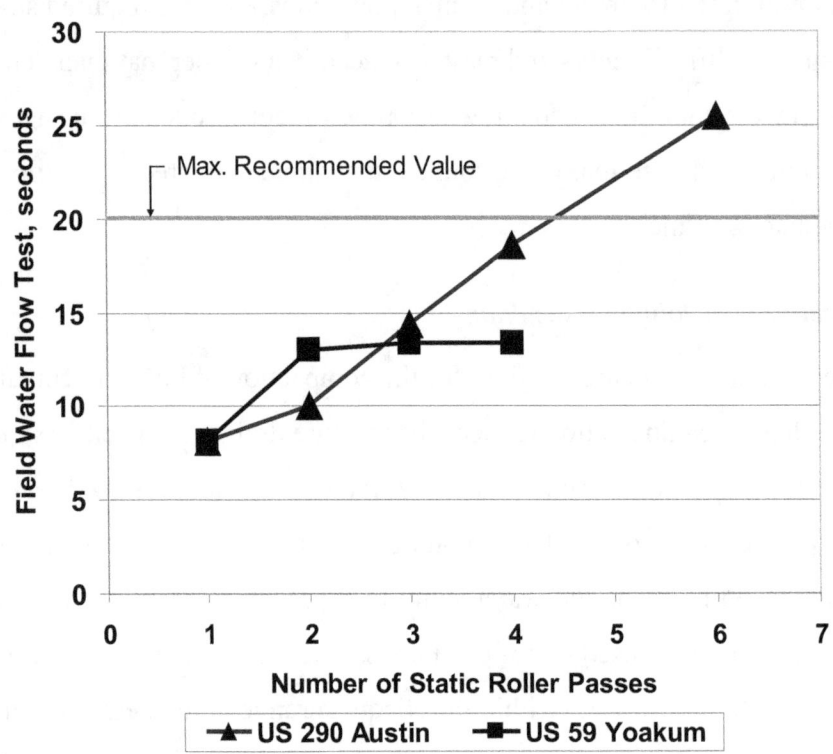

Figure 95. Field Water Flow Value versus Field Compaction Effort.

Table 27. Mixture Designs Used on US 59 Yoakum and US 290 Austin.

Sieve Size	US 59 Yoakum Mix, % pass	US 290 Austin Mix, % pass
3/4 inch	100.0	100.0
1/2 inch	84.5	99.7
3/8 inch	52.8	75.7
No. 4	6.6	7.9
No. 8	4.2	1.1
No. 200	2.4	0.6
Binder Type and Content	PG 76-22S w/Fiber 6.0%	Asphalt Rubber 8.3%

Care should also be exercised to minimize the amount of roller overlap that often occurs in the center of the mat. This results in the center of the mat receiving more compaction than the outside edges and is not a problem for dense-graded mixes. Additional compaction due to roller overlap in the center of the PFC mat for the US 290 mixture in Austin (Figure 95) could restrict the lateral flow of water through the mix.

One of the requirements implemented by the Houston District is that only one roller pass is allowed (i.e., rollers are not permitted to back up). This requires two rollers operating in tandem, at a very slow speed, to achieve one full coverage of the mat (Figure 96). Water flow values for this pavement after one roller pass was about 10 seconds. Stopping the roller on the mat for an extended length of time may leave a roller mark.

Figure 96. A Single Roller Pass as Allowed in the Houston District (No Roller Back Up Allowed).

Longitudinal joints should always be located outside the wheel paths. Most districts use a conventional butt joint, though the Yoakum District has successfully constructed the notched wedge joint as shown previously in Figure 94. Longitudinal joints for PFCs should be constructed in much the same manner as for dense-graded mixes.

Care should be taken not to pinch the outside edge of the mat during compaction, which could impede the flow of water to the shoulder. Joint adhesives or tack coats are sometimes placed on the longitudinal unsupported mat edge to improve the bond to the subsequent lane at the joint interface of dense-graded mixtures. This practice should be avoided for PFC mixtures since additional binder at the joint interface could reduce the permeability and interfere with the lateral movement of water.

Mixture Acceptance

Even though specified density in the field is not currently required, adequate compaction is necessary since low-density zones are prone to raveling. On the other hand, too much compaction can affect the mixture's permeability. The practice in most agencies for mixture approval is based on the evaluation of binder content and gradation and the execution of visual inspection of the mixture after compaction to evaluate (qualitatively but not quantitatively) the density, material variability, and segregation. TxDOT accepts the mixture based on aggregate gradation, lab-molded density, binder content, draindown, boil test, and a thermal profile. In addition, the engineer may take samples or cores from suspect areas to determine recovered asphalt properties. Corrective action is also required if there are any surface irregularities such as segregation, rutting, raveling, flushing, fat spots, mat slippage, color, texture, roller marks, tears, gouges, streaks, or uncoated aggregate particles.

Forensic investigations of raveling failures associated with PFCs in Texas have revealed that the raveled areas had lower asphalt contents than other portions of the project. Quality control of the asphalt content is an important factor in ensuring long life.

MAINTENANCE

PFC mixtures may exhibit the following distress modes:
- Shear failures in high stress areas.
- Cracking due to fatigue.
- Cracking due to reflection from below.
- Raveling due to oxidation and hardening of the binder.
- Raveling due to softened binder from oil and fuel drippings.
- Raveling due to lack of compaction or low asphalt content.
- Delamination due to improper tack coat application.
- Clogging of voids from mud or sand causing loss of permeability (a clogged PFC still drains better than a dense graded mix).
- Rich and dry spots due to draindown of binder during transportation and placement.

Most TxDOT PFC mixtures that are designed and placed under Item 342 have performed well with little to no maintenance required. No rutting has been observed on any of the in-place mixes throughout this research project, and many of the PFCs are under very heavy traffic. Some minimal cracking has been observed, which appears to be a reflection of underlying cracks. Longer-term performance concerns for PFCs are with regard to raveling and delamination, and this has occurred on only a few PFCs in the state. One of the original Item 342 PFCs was placed on IH-35 just north of San Antonio. This mixture started to exhibit some isolated performance problems at about 4 years of age. Small isolated areas in the wheel paths were exhibiting signs of delamination or raveling or both.

TTI took some cores from this roadway in 2007, and these cores are shown in Figure 97. This mix was constructed with 50 percent sandstone, 50 percent limestone, and an asphalt rubber binder. The pictures in Figure 97 show the bottom of the PFC layer. There was no seal coat under the PFC. The failures on this PFC were occurring in the wheel paths. Note the underside of the core taken from the wheel path in Figure 97. The tack coat no longer seems to be functioning, and in fact, the cleanliness of the aggregate surfaces indicates the asphalt binder from the mix as well as the tack may have stripped from the aggregate surfaces.

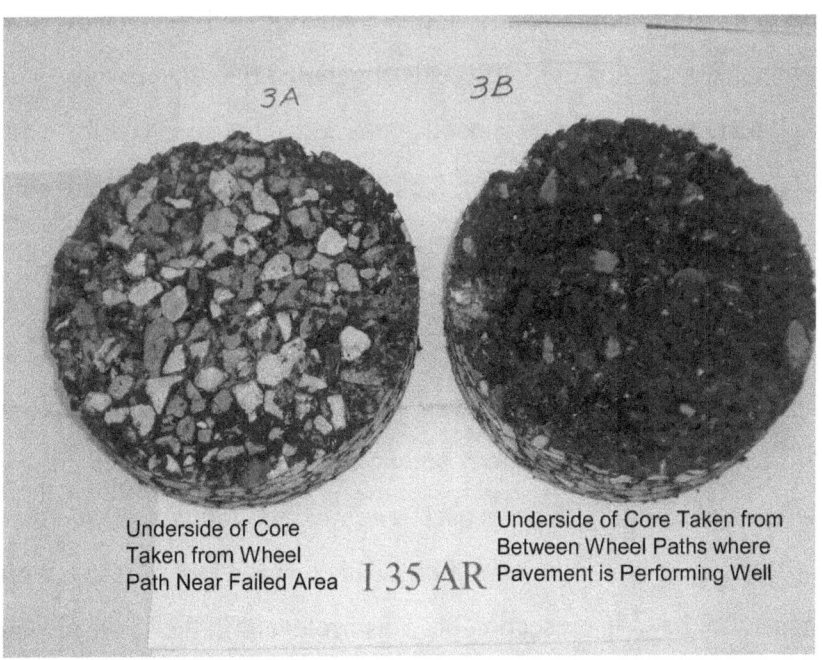

Figure 97. Cores Taken from Distressed PFC on IH-35 in San Antonio.

Because these were small isolated areas, the maintenance section was able to repair the mix with dense graded patching materials without severely impacting the drainage characteristics of the mix.

When the area to be repaired is small and the flow around the patch can be ensured, dense-graded mix is recommended for patching. Otherwise, the zone should be repaired by using PFC mixture. To diminish the wheel impact on the patch joint and facilitate the flow of water around a dense graded patch, rotation of the patch to 45 degrees to provide a diamond shape is recommended.

According to a survey conducted as part of the National Cooperative Highway Research Program (NCHRP) Synthesis 284, there are no reports in the United States on the application of major maintenance for PFC. From 17 states that reported their use, only New Mexico, Wyoming, South Carolina, and Oregon employ fog seals to perform preventive maintenance. Although quantitative information about the significance of these treatments is not available, it is expected that fog seals extend the life of porous mixtures since they provide a small film of unaged asphalt at the surface. FHWA recommends fog seal application in two passes (at a rate of 0.05 gal/yd^2 for each pass) using a 50 percent dilution of asphalt emulsion without any rejuvenating agents (*40*).

Research in Oregon regarding permeability reduction and changes in pavement friction on certain PFC pavements generated by fog seals concluded that the mixtures still retain porosity and keep the rough texture related to its capability to reduce the potential for hydroplaning (*22*). However, quantitative conclusions regarding the changes in these parameters are not included. A decrease in pavement friction was noticed immediately after fog seal application, but during the first month, it increased considerably by traffic action.

Rhode Island further reported that recessed thermoplastic traffic markings proved cost effective in comparison with non-recessed thermoplastic markings. Although recessed thermoplastic traffic markings showed lower snowplow blade damage, fully and semi-recessed markings installed in a tangent highway test section failed to maintain the recommended minimum retroreflectivity in wet night conditions. This result was associated with the effect of the water film present in the tangent section but was irrelevant in the super-elevated curved test section included in the research (*41*).

Highway agencies in British Columbia, South Carolina, and Maryland reported that thermoplastic marking material was the most appropriate for PFC applications (*21*). The British limit the use of pavement markings with thermoplastic materials to certain directional signs and arrows, considering that in PFC the marking material has more opportunity to flow downward into the mixture (*42*). Although higher demand of marking material in PFC (due to higher porosity) was reported by some agencies in the United States (e.g., Ohio, New York, and Oregon), there were no specific recommendations regarding materials for traffic marking (*21*).

In general, PFC mixtures exhibit lower thermal conductivity and reduced heat capacity compared with dense-graded hot mix. Elevated air voids contents in PFC reduce the flow rate of heat through the material. In fact, the thermal conductivity of PFC can be 40 to 70 percent the magnitude of that for dense-graded mix, making PFC operate as an insulating course at the surface.

As a result of these thermal properties, the surface of a PFC can exhibit temperatures 2 to 4°F lower than the surface temperature of adjacent dense-graded mix, producing earlier and more frequent frost and ice formation. Longer periods under such conditions, compared with dense graded mixes, are thus expected. Thus, the time to reach adequate pavement friction values after ice formation has occurred is longer in porous pavement. In fact, formation of black ice and extended frozen periods are currently considered the main problems associated with PFC maintenance in the United States.

In Texas, severe weather events are generally confined to the northern section of the state. It is in these areas that district personnel must prepare for winter maintenance strategies for PFC pavements.

As is indicated from the literature and the current practice of TxDOT districts, anti-icing procedures may produce the best result to combat black ice, freezing rain, and light snow events (*11*). Anti-icing procedures involve a combination of liquid, dry solid, and prewetted chemicals applied at the appropriate times, taking into consideration temperature, the amount of moisture and traffic conditions. De-icing procedures should be reserved for events in which ice and snow have already bonded. These procedures generally require more materials and do not maintain safe road conditions as well as anti-icing procedures.

Sand should only be used in emergency situations where quick friction is needed, for instance, during a surprise ice or snow event (*11*). Use of sand on these pavements may cause clogging to occur, which reduces the draining benefits of PFC. The use of other materials may be used to generate the needed friction. Table 28 shows a plan for anti-icing and de-icing operations suggested by the FHWA in a black ice event (43).

Table 28. Weather Event: Frost or Black Ice (11,43).

Pavement Temp. Range and Trend and Relation to Dew Point	Traffic Condition	Initial Operation		Subsequent Operations		Comments
		Maint. action	Dry chemical spread rate, lb/lane-mi	Maint. action	Dry chemical spread rate, lb/lane-mi	
			liquid / solid or prewetted solid		liquid / solid or prewetted solid	
Above 32°F, steady or rising	Any level	None, see comments		None, see comments		Monitor pavement temperature closely; begin treatment if temperature starts to fall to 32 F or below and is at or below dew point
28 to 35°F, remaining in range or falling to 32 F	Traffic rate less than 100 vehicles per h	Apply prewetted solid chemical	25-65 (solid)	Reapply prewetted solid chemical as needed	25-65 (solid)	1) Monitor pavement closely; if pavement becomes wet or if thin ice forms, reapply chemical at higher indicated rate
2) Do not apply liquid chemical on ice so thick that the pavement cannot be seen
32 F or below, *and* equal to or below dew point	Traffic rate greater than 100 vehicles per h	Apply liquid or prewetted solid chemical	25-65 / 25-65	Reapply liquid or prewetted solid chemical as needed	40-115 / 25-65	
20 to 28°F, remaining in range, *and* equal to or below dew point	Any level	Apply liquid or prewetted solid chemical	65-130 / 65-130	Reapply liquid or prewetted solid chemical when needed	65-130 / 65-130	1) Monitor pavement closely; if thin ice forms, reapply chemical at higher indicated rate
2) Applications will need to be more frequent at higher levels of condensation; if traffic volumes are not enough to disperse condensation, it may be necessary to increase frequency
3) It is not advisable to apply a liquid chemical at the indicated spread rate when the pavement temperature drops below 23 F
15 to 20°F, remaining in range, *and* equal to or below dew point	Any level	Apply prewetted solid chemical	130-200 (solid)	Reapply prewetted solid chemical when needed	130-200 (solid)	1) Monitor pavement closely; if thin ice forms, reapply chemical at higher indicated rate
2) Applications will need to be more frequent at higher levels of condensation; if traffic volumes are not enough to disperse condensation, it may be necessary to increase frequency
Below 15°F, steady or falling	Any level	Apply abrasives		Apply abrasives as needed		It is not recommended that chemicals be applied in this temperature range

Notes:
Timing: (1) Conduct initial operation in advance of freezing. Apply liquid chemical up to 3 hrs in advance. Use longer advance times in this range to effect drying when traffic volume is low. Apply prewetted solid 1 to 2 hrs in advance. (2) in the absence of precipitation, liquid chemical at 75 lb/lane-mi has been successful in preventing bridge deck icing when placed up to 4 days before freezing on higher volume roads and 7 days before on lower volume roads.

CHAPTER 7:
SUMMARY

The objective of this project was to evaluate the performance of porous friction course pavements (PFCs) over time, quantify the costs and benefits of these pavements, and compare them to other wearing surface pavement types. Field performance of several pavement sections from previous TxDOT projects and new construction including Asphalt Rubber (AR) PFCs, Performance Graded (PG) PFCs, thin-bonded PFCs (TBPFC), Ultra Thin Bonded Hot Mix Wearing Courses (UTBHMWC), and dense-graded Hot Mix Asphalt (HMA) were monitored over a four-year period. Non-destructive on-site tests included noise measurements using On Board Sound Intensity (OBSI) measurement equipment, drainability following the Tex-246-F field water flow test procedure, Mean Profile Depth (MPD) using a circular texture meter (CTM), friction at 60 kph using a Dynamic Friction Tester (DFT), and skid that TxDOT collected using a skid trailer. The change of these performance variables with time as well as the influence of traffic, binder/mixture type, aggregate classification, and climatic region was evaluated. In addition, a visual assessment of the condition of the pavement to identify raveling, rutting, or cracking distress was also performed. Accident data on a comprehensive number of PFC sections in the state were also gathered and analyzed. All this information was compiled in database format and made available to TxDOT personnel. Besides the on-site measurements, when performance issues were identified on the PFC sections, field cores were acquired and a laboratory forensic evaluation conducted to identify the possible causes of pavement distress. Results from the multiyear performance data analysis and previous research were used to produce guidelines and recommendations to improve the design, construction, and maintenance of PFCs.

Performance evaluation of PFCs over time demonstrated adequate performance. There was a slight increase in OBSI noise levels with time, but the correlation was weak. The trend was observed only when all measurements were combined. When each pavement section was individually analyzed, no particular trend was observed, except for sections 9 and 14 where a substantial increase in noise levels was noted due to raveling. The type of mixture had a significant effect in noise levels. PFCs had lower overall noise levels when compared to dense-graded HMA reference sections, although not statistically significant. Binder type was also a significant factor in the acoustic performance of PFCs, with the statistical analysis showing significant difference between the AR-PFCs (quieter) vs. the PG-PFCs. The climatic region proved a significant factor as well, as the statistical analysis indicated that there was a significant difference in noise levels with respect to each section's climatic region.

For drainability, PG-PFCs showed better performance as compared to AR-PFCs. In general, measurements in AR-PFCs were more variable than PG-PFCs. For PG-PFCs, the water flow values (WFVs) had a tendency to increase early in the life of the pavement and remain relatively constant afterwards. A 90-second threshold was established to define pervious vs. impervious pavement sections after comparing the WFVs acquired on wheel path (OWP) vs. between wheel path (BWP). All PG-PFCs were below the set threshold, while most of the AR-PFCs were close to or above the 90-second limit. The classification of the aggregates per the Surface Aggregate Classification (SAC) system did not play a role with respect to WFVs. On the other hand, climate had a significant role. Apparently, the amount of rainfall in a particular climatic region helped assure the continued drainability of PFCs, especially in warm climates.

With regard to the texture measurements, there was a strong correlation between the MPD values acquired OWP and BWP, with texture measured BWP being slightly larger than texture measured OWP. With time, the texture values remained practically unchanged. The two exceptions, where texture changed significantly, corresponded to a pavement section that was fog sealed and the other one that exhibited significant amounts of raveling. Both AR and PG-PFCs had statistical equivalent texture and superior than the dense-graded HMA pavement sections used as reference. Neither the aggregate classification nor the climatic region where the pavement sections were placed had a significant impact on texture. Pavement texture achieved soon after construction seemed to be more critical to performance over time.

Friction measurement acquired OWP vs. BWP also showed a strong correlation, with BWP friction values being slightly higher than OWP, especially for non-PFC pavements. With time, PG-PFCs demonstrated an upward trend in friction values, while AR-PFCs and non-PFC pavement sections show a flat or downward friction trend. Differences in pavement type did not have an influence on friction. Aggregate classification, however, did play an important role. Pavement sections with SAC-B aggregates had statistically significantly lower friction values as compared to those pavement sections employing either SAC-A or SAC-A/B aggregates. The effect of climatic region on friction demonstrated that pavement sections in dry climates had lower friction values while pavement sections located in the wet cold climatic region had the highest friction.

With respect to skid number (SN) measurements, these seemed more variable and did not follow a consistent upward or downward trend with time. Similar to the texture observations, AR-PFCs and PG-PFCs had higher SN values compared to dense-graded HMA. Moreover, in line with the friction results, the SN for mixtures with different aggregate classifications was statistically different. Sections employing SAC-A aggregates had the highest SNs, while sections where mixtures were comprised of only SAC-B aggregates had the lowest SNs. Pavement sections in the DW climatic region had the lowest SNs, while pavement sections located in the WC regions had the highest SNs. The International Friction Index (IFI) was calculated using texture and friction measurements, and a correlation between IFI and SN was developed.

The analysis to investigate the safety of PFC under wet conditions included 161 PFC sections from across Texas constructed between 2003 and 2011. Accident rates on these sections were compared before and after construction of the PFC surface. To investigate the safety of PFC under wet conditions, the weather and surface conditions at the time of an accident as reported in CRIS was used but found to provide inconsistent and inconclusive results. To better investigate the influence of rain conditions on the safety of PFCs, the research team used 280 weather stations around Texas to estimate the rainfall totals on the day of an accident. By evaluating only the subset of accidents with daily rainfall estimates greater than 2.5 mm (0.1 inch), the researchers obtained a clearer distinction in the wet weather safety of PFCs. The data indicate that PFC indeed reduces the number of accidents, injuries, and fatalities on roads in Texas. Using the accident data as reported in CRIS, a slight but consistent increase in accidents on PFC was observed immediately following construction of these surfaces. Therefore, it is recommended that the initial safety of PFC be addressed before opening the road to traffic. The CRIS database should be updated to include rainfall totals on the day of an accident to better investigate and improve the safety of roads in wet conditions.

The Generalized Benefit/Cost Ratio (GBC) methodology was used successfully in this project despite the limited dataset (i.e., 17 pavement sections). The estimation of the B/C ratio was based on four variables (attributes) that were statistically significant at the 95 percent level. Furthermore, the model fitted the observations well with an R-squared of 0.87. The model could be further improved by incorporating information from reference and additional PFC sections.

All sections monitored as part of this project are relatively new and are in relatively good condition. It is recommended that TxDOT continue to monitor all PFC pavement sections evaluated as part of this research project until maintenance or rehabilitation of the PFC is performed. Only then, actual performance can be incorporated into the B/C model.

Based on the findings of this project, the continued use of PFC in Texas is strongly encouraged. Specific considerations with regard to usage, cost, materials and design, construction, and maintenance are provided in Chapter 6.

REFERENCES

1. Texas Department of Transportation (TxDOT), *Standard Specifications for Construction and Maintenance of Highways, Streets, and Bridges Adopted by the Texas Department of Transportation*, Austin, Texas, 2004.
2. Brown, J. R., *Pervious Bitumen–Macadam Surfacings Laid to Reduce Splash and Spray at Stonebridge, Warwickshire*, Report No. LR 563, Transportation Road Research Laboratory, England, 1973.
3. Button, J. W., E. G. Fernando, and D. R. Middleton, *Synthesis of Pavement Issues Related to High-Speed Corridors*, Report No. 0-4756-1, Texas Transportation Institute, Texas A&M University, College Station, Texas, 2004.
4. Kearfott, P., M. Barrett, and J. F. Malina, Jr., *Stormwater Quality Documentation of Roadside Shoulders Borrow Ditches,* CRWR Online Report 05-02, Center for Research in Water Resources, The University of Texas at Austin, Austin, Texas, 2005. http://www.crwr.utexas.edu/online.shtml
5. Khalid, H. and F. Perez, "Performance and Durability of Bituminous Materials", *Performance Assessment of Spanish and British Porous Asphalts*, E & FN Spon, London, pp. 137-157, 1996.
6. Ruiz, A., R. Alberola, F. Perez, and B. Sanchez, *Porous Asphalt Mixtures in Spain*, Transportation Research Record 1265, pp. 87–94, 1990.
7. Rand, D., "Permeable Friction Courses (PFC) in Texas", *Presentation for the Annual School for County Commissioners Courts,* College Station, TX, February 8, 2012.
8. Cooley Jr., L. A., R. B. Mallick, W. S. Mogawer, M. Partl, L. Poulikakos, and G. Hicks, *Construction and Maintenance Practices for Permeable Friction Courses,* NCHRP Report 640, National Cooperative Highway Research Program, Transportation Research Board, Washington, DC, 2009.
9. Alvarez, A. E., A. Epps Martin, C. Estakhri, J. W. Button, C. Glover, and S.H. Jung, *Synthesis of Current Practice on the Design, Construction, and Maintenance of Porous Friction Courses*, Report No. FHWA/TX-06/0-5262-1, Texas Transportation Institute, College Station, TX, 2006.
10. Trevino, M., and T. Dossey, *Noise Measurements of Highway Pavements in Texas*, Research Report 0-5185-3, Center for Transportation Research, The University of Texas at Austin, October 2009.
11. Yildirim, Y., T. Dossey, K. Fults, M. Tahmoressi, and M. Trevino, *Winter Maintenance Issues Associated with New Generation of Open-Graded Friction Courses*, Report No. FHWA/TX-08-0-4834-2, Center for Transportation Research, Austin, TX, February 2007.
12. Liu, K-W., A.E. Alvarez, A. Epps Martin, T. Dossey, A. Smit, C.K. Estakhri, *Synthesis of Current Research on Permeable Friction Courses: Performance, Design, Construction, and Maintenance*, FHWA/TX-10/0-5836-1, Texas Transportation Institute, February 2010.
13. Donavan, P. R., *Comparative Measurement of Tire/Pavement Noise in Europe and the United States*, Illingworth & Rodkin, Inc., 2005.
14. Alvarez, A. E., A. Epps Martin, and C. Estakhri, *Drainability of Permeable Friction Course Mixtures*, Journal of Materials in Civil Engineering, Vol. 22, No. 6, pp. 556-564, 2010.

15. Alvarez, A. E., A. Epps Martin, C. Estakhri, and R. Izzo, *Evaluation of Durability Tests for Permeable Friction Course Mixtures*, International Journal of Pavement Engineering, Vol. 11, No. 1, pp. 49-60, 2010.
16. Molenaar, A. A. A., A. J. J. Meerkerk, M. Miradi and T. van der Steen, *Performance of Porous Asphalt Concrete*, Proceedings of the Annual Meeting and Technical Sessions, Association of Asphalt Paving Technologist, Savannah, Georgia, CD-ROM, Vol. 75, 2006.
17. Luce, A., E. Mahmoud, E. Masad, and A. Chowdhury, *Relationship of Aggregate Microtexture to Asphalt Pavement Skid Resistance*, Journal of Testing and Evaluation, Vol. 35, No. 6, 2007.
18. Miller, M. M. and H. D. Johnson, *Effects of Resistance to Skidding on Accidents: Surface Dressing on an Elevated Section of the M4 Motorway*, Report No. LR 542, Transport and Road Research Laboratory, Berkshire, United Kingdom, 1973.
19. Kamel, N. and T. Gartshore, *Ontario's Wet Pavement Accident Reduction Program*, ASTM Special Technical Publication 763, American Society of Testing and Materials, Philadelphia, Pennsylvania, 1982.
20. Alvarez, A. E., A. Epps Martin, C. K. Estakhri, *Optimizing the Design of Permeable Friction Course Mixtures*, Transportation Research Record: Journal of the Transportation Research Board, Issue No. 2209, pp. 26–33, 2011.
21. Huber, G., *Performance Survey on Open-Graded Friction Course Mixes*, Synthesis of Highway Practice 284, TRB, National Research Council, Washington., D.C., 2000.
22. Rogge, D., *Development of Maintenance Practices for Oregon F-Mix*, Report No. FHWAOR-RD-02-09. Federal Highway Administration, U. S. Department of Transportation, Washington, D.C., 2002.
23. Donovan, P., and B. Rymer, *Quantification of Tire/Pavement Noise: Application of the Sound Intensity Method*, Presented at the 33rd International Congress and Exposition on Noise Control Engineering, Prague, Czech Republic, August 22–25, 2004.
24. American Association of State Highway and Transportation Officials (AAHSTO), *Standard Method of Test for Measurement of Tire/Pavement Noise Using the On-Board Sound Intensity (OBSI) Method*. AASHTO Designation: TP 76-13, Washington, DC.
25. Trevino, M., and T. Dossey, *A Research Plan for Measuring Noise Levels in Highway Pavements in Texas*, Research Report 0-5185-1, Center for Transportation Research, The University of Texas at Austin, November 2006.
26. Trevino-Frias, M., and T. Dossey, *Preliminary Findings from Noise Testing on PFC Pavements in Texas*, Research Report 0-5185-2, Center for Transportation Research, The University of Texas at Austin, April 2007.
27. American Society for Testing and Materials (ASTM), *Standard Specification for P225/60R16 97S Radial Standard Reference Test Tire*, ASTM Standard Designation: F2493-08. ASTM International, West Conshohocken, PA.
28. Texas Department of Transportation (TxDOT), *Permeability or Water Flow of Hot Mix Asphalt*, Tex-246-F, Construction Division, November 2009.
29. Henry, J., *Evaluation of Pavement Friction Characteristics: A Synthesis of Highway Practice*, NCHRP Synthesis 291, National Cooperative Highway Research Program, 2000.
30. Yeaman, J, *Are We Afraid on the IFI?*, Proceedings of the International Conference on Surface Friction, Christchurch, New Zealand, May 1-4, 2005.

31. Jayawickrama, P. W., R. Prasanna, and S. P. Senadheera, *Survey of State Practices to Control Skid Resistance on Hot-Mix Asphalt Concrete Pavements*, Transportation Research Record: Journal of the Transportation Research Board, No. 1536, pp. 52-58, 1996.
32. Smit, A, J. Prozzi, and, A. Bianchini, *Evaluation of the OBSI Method*, Transportation Research Board Paper 10-3599, TRB 2010 Annual Meeting, CD-ROM.
33. NOAA, *Texas Precipitation*, National Oceanic and Atmospheric Administration. http://www.ncdc.noaa.gov, 2012.
34. NHTSA, *Fatality Analysis Reporting System (FARS)*, National Highway Traffic Safety Administration. http://www.nhtsa.gov, 2012.
35. NOAA, Global Surface Summary of Day, National Oceanic and Atmospheric Administration, http://www.ncdc.noaa.gov, 2012.
36. Van Der Zwan, *Developing Porous Asphalt For Freeways In The Netherlands: Reducing Noise, Improving Safety, Increasing Service Life*, TR News 272, Transportation Research Board, January-February 2011.
37. Elvik, R, and G. Poul, *Safety Aspects Related to Low Noise Road Surfaces*, TOI Report No. 680/2003, Institute of Transport Economics, Oslo, November 2003.
38. Newcomb, D., and L. Scofield, *Quiet Pavements Raise the Roof in Europe: Scanning Tour Reveals European Practice for Noise Mitigation*, Materials and Mixes, September/October 2004.
39. Cooley Jr., L.A., *Annotated Literature Review for NCHRP Report 640*, NCHRP Web-Only Document 138, National Cooperative Highway Research Program, January 2009.
40. Federal Highway Administration (FHWA), *Open-Grade Friction Courses FHWA Mix Design Method*, Technical Advisory T 5040.31, Federal Highway Administration, U.S. Department of Transportation, Washington D.C., 1990.
41. Lee, W., S. Cardi, and S. Corrigan, *Implementation and Evaluation of Traffic Marking Recesses for the Application of Thermoplastic pavement markings on Modified Open Graded Friction Course*, University of Rhode Island – Sponsoring agency: New England Transportation Consortium, Storrs, CT, 1999.
42. The Highways Agency, The Scottish Office Development Department, The Welsh Office Swyddfa Gymreig, The Department of the Environment for Northern Ireland, *Design Manual for Roads and Bridge, Volume 7: Pavement Design and Maintenance Bituminous Surfacing Materials and Techniques*, 1999.
43. Federal Highway Administration (FHWA), *Manual of Practice for an Effective Anti-Icing Program: A Guide for Highway Winter Maintenance Personnel*, U.S. Army Cold Regions Research and Engineering Laboratory Corps of Engineers, Hanover, New Hampshire.

APPENDIX A:
PAVEMENT DATABASE

A product of this research study was a database comprising PFC projects and other type of pavements evaluated as part of the study. The database includes project information as well as material and specific test data collected during the study. This appendix provides an overview of the tables included in the database and a description of the various data fields.

Table 1 provides a list of the projects for which data were collected as part of the study. Each project was given a unique project ID that was used in the database tables to indicate the project for which data are reported. In addition to PFC projects, the list also includes dense-graded HMA reference sections (IDs: 10, 13, 16 and 18) and Ultra Thin Bonded Hot Mix Wearing Course (ID: 8, 23, 24, and 25). These sections were identified for comparative purposes.

The data tables expand on the properties of these projects although the performance tests completed as part of the study were not done on each of the projects listed—some projects, for example, were evaluated as case studies.

DATABASE TABLES

The PFC database comprises the following data tables. A brief overview of each is provided in Table A.1.

Table A.1. Database Tables.

Database table	Overview
AGG_GRAD	Aggregate gradation and Cantabro loss data
BNDR	Binder type and binder content information (as well as filler and additive information)
CRASH	Accident data collected to evaluate wet weather safety
CTM	Circular texture meter (CTM) surface macrotexture
CTM_DROP	Drop-outs (inaccurate) CTM measures
CTM_MPD	Mean profile depth CTM measures
CTM_RMS	Root mean square CTM measures
DFT	Dynamic friction test (DFT) measurements
OBSI	On-board sound intensity (OBSI) measurements
PERFORMANCE	Visual observation of performance (rutting, raveling, etc.)
PRJ	Project specific information such as location, longitudes, latitudes, etc.
RUT	Straight-edge rutting measurements
SKID_TRAFFIC	Skid measurements using TxDOT skid trailer
WFV	Water flow values (WFV)

AGG_GRAD

This table includes aggregate gradations of the PFC project mixes as well as Cantabro loss values as reported in the mix design or QCQA sheets (from SiteManager). Table A.2 provides the fields and descriptions.

Table A.2. AGG_GRAD Data Field Descriptions.

Field	Description
ID	Unique record identifier
PRJ_NBR	Project number based on 5836 ID
P3/4	Percentage passing 3/4" sieve
P5/8	Percentage passing 5/8" sieve
P1/2	Percentage passing 1/2" sieve
P3/8	Percentage passing 3/8" sieve
P4	Percentage passing #4 sieve
P8	Percentage passing #8 sieve
P10	Percentage passing #10 sieve
P16	Percentage passing #16 sieve
P30	Percentage passing #30 sieve
P 40	Percentage passing #40 sieve
P50	Percentage passing #50 sieve
P 80	Percentage passing #80 sieve
P100	Percentage passing #100 sieve
P200	Percentage passing #200 sieve
CANT	Cantabro loss as reported in mix design or QC/QA reports

BNDR

This table includes information on the binder grade and content for the project mixes including types of modifiers, fiber content, and use of additives. Table A.3 provides the fields and descriptions.

Table A.3. BNDR Data Field Descriptions.

Field	Description
ID	Unique record identifier
PRJ_NBR	Project number based on 5836 ID
P3/4	Percentage passing 3/4" sieve
P5/8	Percentage passing 5/8" sieve
P1/2	Percentage passing 1/2" sieve
P3/8	Percentage passing 3/8" sieve
P4	Percentage passing #4 sieve

CRASH

This table includes information extracted from TxDOT's Crash Records Information System (CRIS) between 2003 and 2011 for the PFC project sections. Only a subset of the data fields available in CRIS were extracted for an analysis of PFC wet weather safety. Each of the records in the table represents a reported crash. Further information for these crashes may be obtained by linking the Crash_ID field in this table to CRIS. The CRIS database provides lookup tables for various identifier fields. Tables A.5 through A-11 provide a summary of the identifier fields in the crash database table.

Table A.4. CRASH Data Field Descriptions.

Field	Description
YR	Year when the crash was reported
PRJ_NBR	Project number based on 5836 ID
Crash_ID	Unique CRIS crash ID
Crash_Date	Date of crash
Crash_Time	Time of crash
Lat	Latitude of crash
Long	Longitude of crash
Road_Name	Road on which crash occurred
RM	Texas reference marker at which crash occurred
Wthr_Cond_ID	See Table 6
Light_Cond_ID	See Table 7
Road_Algn_ID	See Table 8
Surf_Cond_ID	See Table 9
Tot_Injry_Cnt	Number of injuries
Death_Cnt	Number of fatalities
Rural_Urban_Type_ID	See Table 10
Crash_Sev_ID	See Table 11
Road_Constr_Zone_Fl	Construction flag
Road_Part_Adj_ID	See Table 12
Nbr_Of_Lane	Number of lanes on road
Construction_Date	Date of construction of PFC section (based on SiteManager acceptance date)

Table A.5. Weather Condition (Wthr_Cond_ID).

ID	Description
0	UNKNOWN
1	CLEAR/CLOUDY
2	RAIN
3	SLEET/HAIL
4	SNOW
5	FOG
6	BLOWING SAND/SNOW
7	SEVERE CROSSWINDS
8	OTHER (EXPLAIN IN NARRATIVE)
9	BLOWING DUST
10	SMOKE
11	CLEAR
12	CLOUDY
94	REPORTED INVALID
95	NOT REPORTED

Table A.6. Light Condition (Light_Cond_ID).

ID	Description
0	UNKNOWN
1	DAYLIGHT
2	DAWN
3	DARK, NOT LIGHTED
4	DARK, LIGHTED
5	DUSK
6	DARK, UNKNOWN LIGHTING
8	OTHER (EXPLAIN IN NARRATIVE)

Table A.7. Road Alignment (Road_Algn_ID).

ID	Description
1	STRAIGHT, LEVEL
2	STRAIGHT, GRADE
3	STRAIGHT, HILLCREST
4	CURVE, LEVEL
5	CURVE, GRADE
6	CURVE, HILLCREST
7	OTHER (EXPLAIN IN NARRATIVE)
8	UNKNOWN
9	NOT REPORTED

Table A.8. Rural-Urban Indicator (Rural_Urban_Type_ID).

ID	Description
-1	NO DATA
1	RURAL (<5000)
2	SMALL URBAN (5000-49,999)
3	LARGE URBAN (50,000-199,999)
4	URBANIZED (200,000+)

Table A.9. Surface Condition (Surf_Cond_ID).

ID	Description
0	UNKNOWN
1	DRY
2	WET
3	STANDING WATER
4	SNOWY/ICY
5	SLUSH
6	ICE
7	MUDDY
8	OTHER (EXPLAIN IN NARRATIVE)
9	SNOW
10	SAND
94	REPORTED INVALID
95	NOT REPORTED

Table A.10. Crash Severity (Crash_Sev_ID).

ID	Description
0	UNKNOWN
1	INCAPACITATING INJURY
2	NON-INCAPACITATING
3	POSSIBLE INJURY
4	FATAL
5	NOT INJURED

Table A.11. Road Crash Location (Road_Part_Adj_ID).

ID	Description
1	MAIN/PROPER LANE
2	SERVICE/FRONTAGE ROAD
3	ENTRANCE/ON RAMP
4	EXIT/OFF RAMP
5	CONNECTOR/FLYOVER
6	DETOUR
7	OTHER (EXPLAIN IN NARRATIVE)
8	TRANSITWAY
9	TRANSITWAY RAMP
10	NOT REPORTED

CTM

The CTM table provides a summary of the circular texture meter macrotexture measurements taken on and between the wheel paths. Multiple sections within each project were tested. Table A.12 provides the fields and descriptions.

Table A.12. CTM Data Field Descriptions.

Field	Description
ID	Unique identifier
FY	Year of data collection
PRJ_NBR	Project number based on 5836 ID
SEC_NBR	Section number
BWP	Average (A-H) MPD between wheel path using CTM
OWP	Average (A-H) MPD measured on wheel path using CTM

CTM_DROP

This table includes information on drop-outs or inaccuracies recorded during the CTM measurements. High dropouts are typical for CTM measurements on PFC surfaces. The measurements A through H correspond to equidistant circular segments across which the CTM measures are reported. Table A.13 provides the fields and descriptions.

Table A.13. CTM DROP Data Field Descriptions.

Field	Description
ID	Unique identifier
FY	Year of data collection
PRJ_NBR	Project number based on 5836 ID
LOC	Location on the measurement: Between Wheel Path (BWP) or On Wheel Path (OWP)
SEC_NBR	Section number
RUN	Run number
A	Dropout Percentage (Measurement A)
B	Dropout Percentage (Measurement B)
C	Dropout Percentage (Measurement C)
D	Dropout Percentage (Measurement D)
E	Dropout Percentage (Measurement E)
F	Dropout Percentage (Measurement F)
G	Dropout Percentage (Measurement G)
H	Dropout Percentage (Measurement H)

CTM_MPD

The CTM_MPD table provides the raw data used to calculate the average summaries reported in Table MPD. Macrotexture measurements at each of the equidistant circular segments A–H are provided. Table A.14 provides the fields and descriptions.

Table A.14. CTM_MPD Data Field Descriptions.

Field	Description
ID	Unique identifier
FY	Year of data collection
PRJ_NBR	Project number based on 5836 ID
SEC_NBR	Section number
LOC	Location on the measurement: Between Wheel Path (BWP) or On Wheel Path (OWP)
RUN	Run number
A	Mean Profile Depth (measurement A)
B	Mean Profile Depth (measurement B)
C	Mean Profile Depth (measurement C)
D	Mean Profile Depth (measurement D)
E	Mean Profile Depth (measurement E)
F	Mean Profile Depth (measurement F)
G	Mean Profile Depth (measurement G)
H	Mean Profile Depth (measurement H)

CTM_RMS

This data table provides the root-mean-square (RMS) data output by the CTM. These data are an alternate measure of macrotexture providing an indication of the nature of the surface texture, i.e., whether positive or negative. Table A.15 provides the fields and descriptions.

DFT

The DFT table provides dynamic friction test (DFT) results. These measurements were taken at the same locations of the CTM measurements. The DFT provides an estimate of the coefficient of wet friction at speeds of 20, 40, 60 and 80 kph. Table A.16 provides the fields and descriptions.

Table A.15. CTM RMS Data Field Descriptions.

Field	Description
ID	Unique identifier
FY	Year of data collection
PRJ_NBR	Project number based on 5836 ID
SEC_NBR	Section number
LOC	Location on the measurement: Between Wheel Path (BWP) or On Wheel Path (OWP)
RUN	Run number
A	Root Mean Square (Measurement A)
B	Root Mean Square (Measurement B)
C	Root Mean Square (Measurement C)
D	Root Mean Square (Measurement D)
E	Root Mean Square (Measurement E)
F	Root Mean Square (Measurement F)
G	Root Mean Square (Measurement G)
H	Root Mean Square (Measurement H)

Table A.16. DFT Data Field Descriptions.

Field	Description
ID	Unique identifier
FY	Year of data collection
PRJ_NBR	Project number based on 5836 ID
SEC_NBR	Section number
BWP20	Between wheel path friction measurement at 20 kph
BWP40	Between wheel path friction measurement at 40 kph
BWP60	Between wheel path friction measurement at 60 kph
BWP80	Between wheel path friction measurement at 80 kph
OWP20	On wheel path friction measurement at 20 kph
OWP40	On wheel path friction measurement at 40 kph
OWP60	On wheel path friction measurement at 60 kph
OWP80	On wheel path friction measurement at 80 kph

OBSI

The OBSI table provides a summary of all on-board sound intensity (OBSI) measurements taken on the test sections. These measurements provide A-weighted noise-levels. The test tire used for OBSI testing was changed during the study as indicated. Table A.17 provides the fields and descriptions.

Table A.17. OBSI Data Field Descriptions.

Field	Description
ID	Unique identifier
FY	Year of data collection
PRJ_NBR	Project number based on 5836 ID
TEST_DATE	Date of test
OBSI_DBA	Mean A-weighted decibel reading
TIRE	Test tire

PERFORMANCE

The performance table provides visual observations of performance for the PFC sections. These observations comment on the condition of the sections during field inspections. Table A.18 provides the fields and descriptions.

Table A.18. PERFORMANCE Data Field Descriptions.

Field	Description
ID	Unique identifier
FY	Year of data collection
PRJ_NBR	Project number based on 5836 ID
OBS	Observations acquired during field inspections

PRJ

The PRJ or project table provides details on the PFC projects including location and extent information as well as dates of field inspections, etc. Table A.19 provides the fields and descriptions.

RUT

The RUT table provides rutting measured in the left and right wheel paths (in the direction of traffic) on the test sections using a straight-edge. Table A.20 provides the fields and descriptions.

SKID_TRAFFIC

This table provides skid numbers on the sections measured by TxDOT using the skid trailer with a smooth test tire at a speed of 50 mph. Table A.21 provides the fields and descriptions.

Table A.19. PFJ Data Field Descriptions.

Field	Description
ID	Unique identifier
PRJ_NBR	Project number based on 5836 ID
CSJ	CSJ number
DISTRICT	District
COUNTY	County
ROUTE	Route number
PMIS ROUTE	PMIS data extraction route
FROM1	Starting point of the project
TO1	Ending point of the project
LAT	Start latitude
LONG	Start longitude
LOC	Project location
DIR	Direction of lane
BTRM	Beginning reference marker
ETRM	Ending reference marker
STRM	Updated Start TRM based on field ops
LANE PMIS	Lane to retrieve data from PMIS (K1, K6, L1, R1, etc.)
PMIS BTRM	PMIS data extraction beginning reference marker
PMIS ETRM	PMIS data extraction ending reference marker
SURF	Type of surface course
BINDER	Binder type
AGG TYPE	Aggregate type
AGG BLEND	Type of aggregate blend
BOND TYPE	Bond type between PFC and the layer below
YEAR 1-DATE	Field inspection date for first round of data collection
YEAR 2-DATE	Field inspection date for second round of data collection
YEAR 3-DATE	Field inspection date for third round of data collection
CONST M/Y	Construction month/year
LET DATE	Let date of CSJ (based on TxDOT Expressway)
CLIMATE	Climate type

Table A.20. RUT Data Field Descriptions.

Field	Description
ID	Unique identifier
FY	Year of data collection
PRJ_NBR	Project number based on 5836 ID
SEC_NBR	Section number
LR	Left wheel path measurement
RR	Right wheel path measurement
Notes	Comments

Table A.21. SKID TRAFFIC Data Field Descriptions.

Field	Description
ID	Unique identifier
PRJ_NBR	Project number based on 5836 ID
FY	Year of data collection
ROUTE	Route number
BTRM	Beginning reference marker number
BDISP	Reference marker displacement
DIST	Distance at measurement point
SN	Skid number
AADT	Annual average daily traffic

WFV

The water flow value (WFV) table provides water flow values measured using the TxDOT falling head permeameter. Measurements were made on and between the wheel paths and in a few instances on the road shoulder. Table A.22 provides the fields and descriptions.

Table A.22. WFV Data Field Descriptions.

Field	Description
ID	Unique identifier
FY	Year of data collection
PRJ_NBR	Project number based on 5836 ID
SEC_NBR	Section number
BWP	Measurement on between wheel path, considered impervious when WFV ≥ 300 sec
OWP	Measurement on wheel path, considered impervious when WFV ≥ 300 sec
SH	Measurement on shoulder; no measurement taken is indicated by "999"

APPENDIX B:
PAVEMENT SECTIONS USED FOR DRAINABILITY, TEXTURE, FRICTION, AND SKID MEASUREMENTS

Appendix B presents the location maps for all 28 fixed and rotating pavement sections indicating the position where the on-site drainability, texture, and friction measurements were performed. The coordinates (i.e., latitude and longitude) for several sections are indicated in the maps.

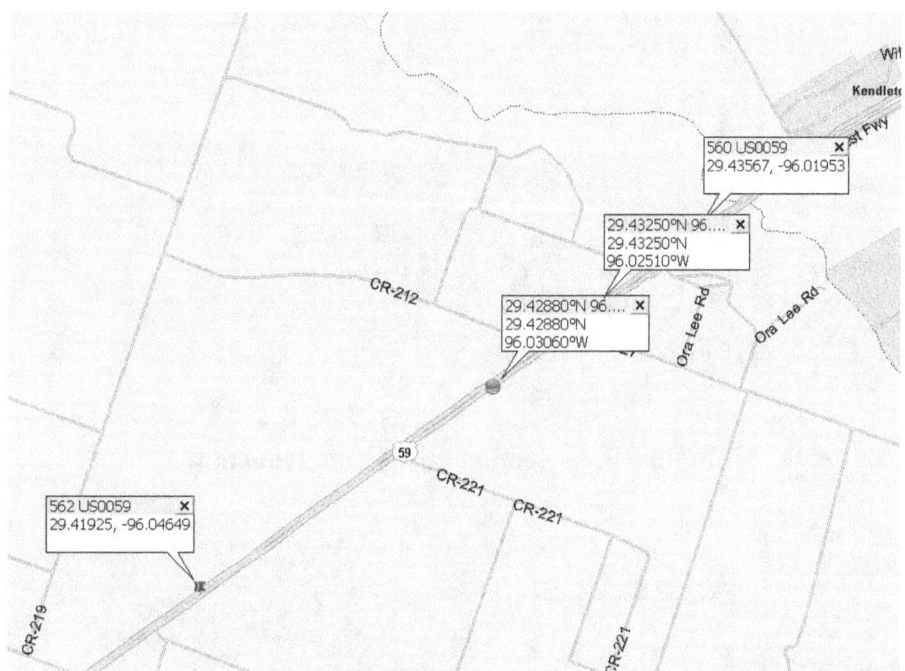

Figure B.1. Section 1 and Section 13, US 59, Yoakum.

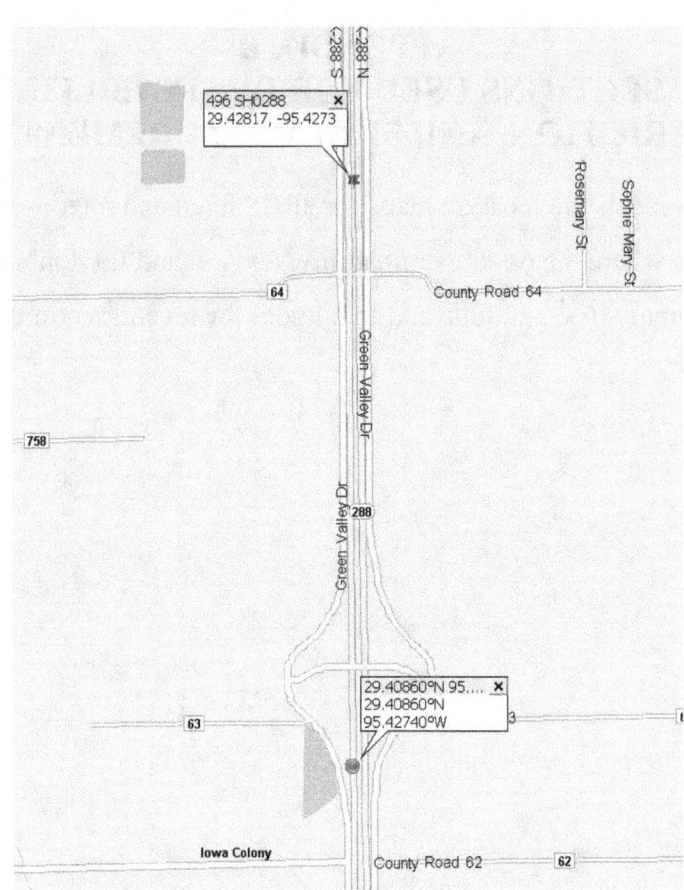

Figure B.2. Section 2, SH 288, Houston.

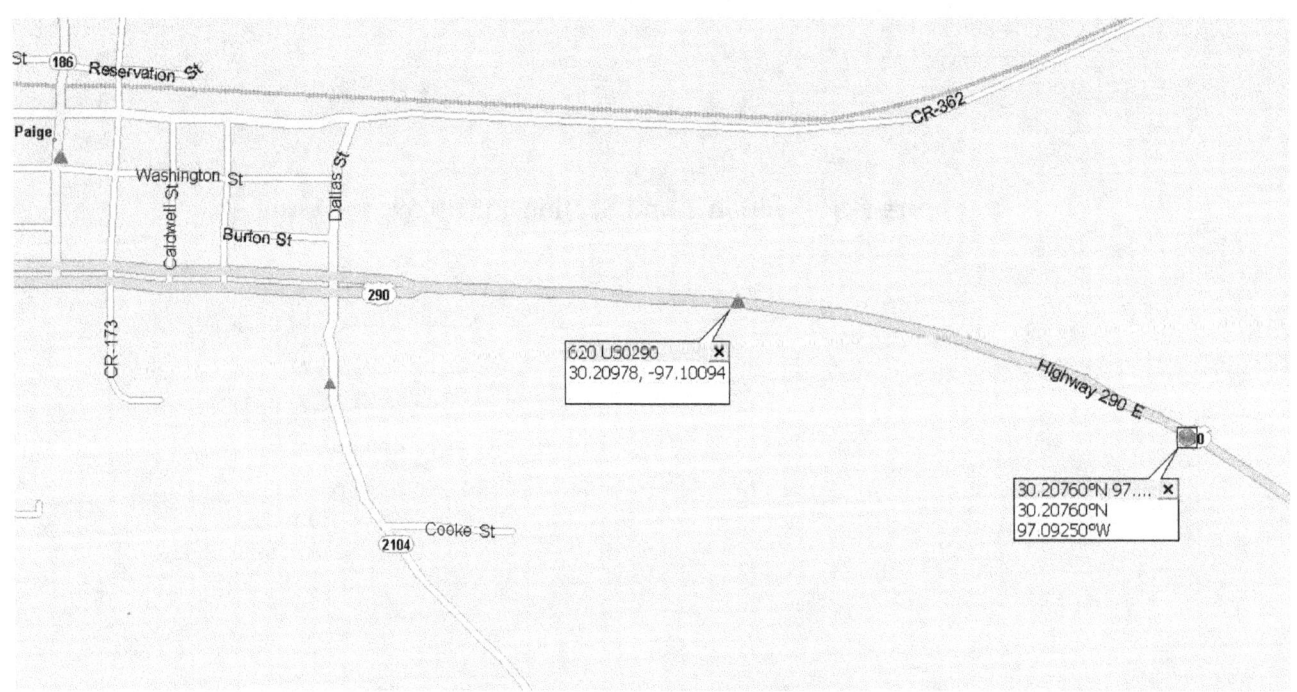

Figure B.3. Section 3, US 290, Austin.

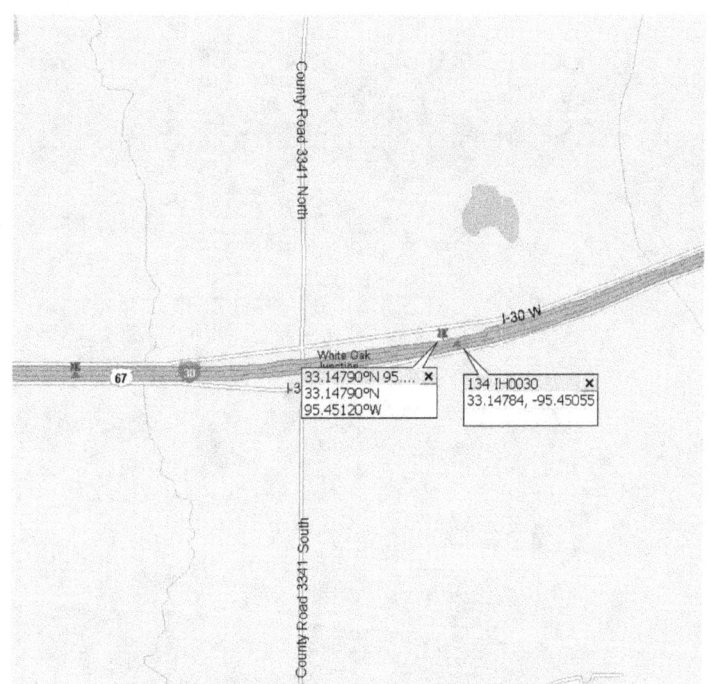

Figure B.4. Section 4, IH 30, Paris.

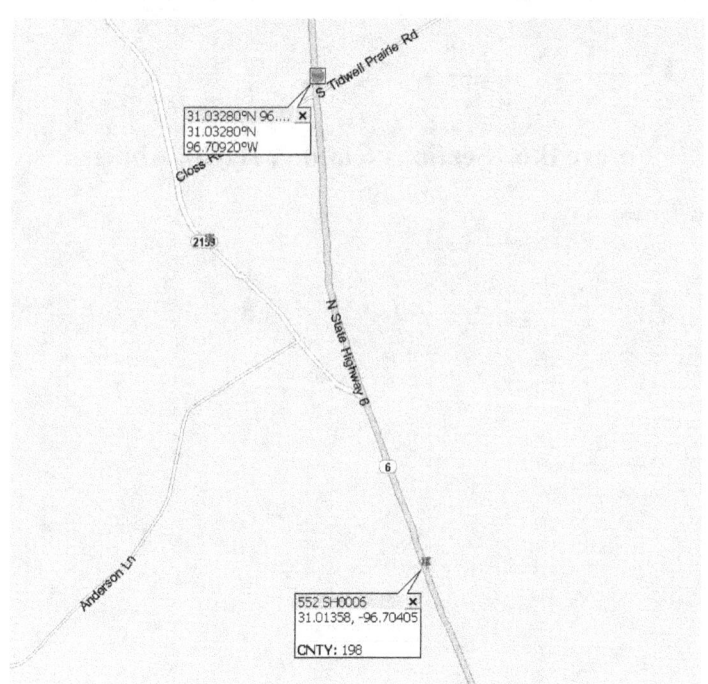

Figure B.5. Section 5, SH 6, Bryan.

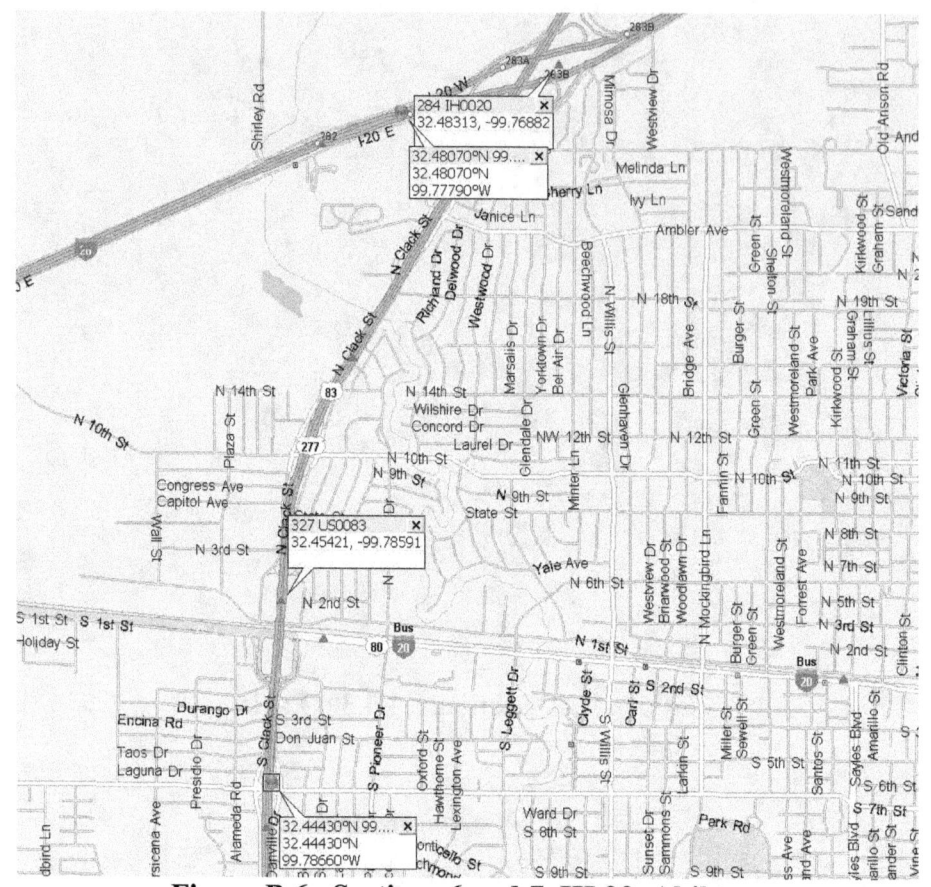

Figure B.6. Sections 6 and 7, IH 20, Abilene.

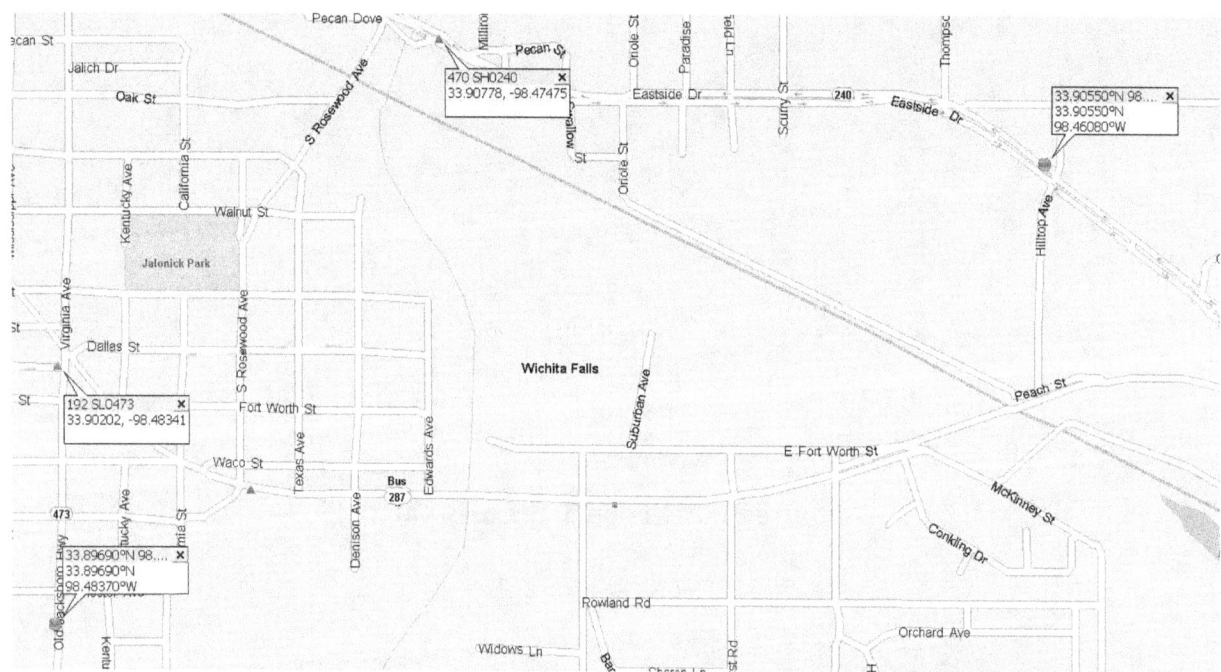

Figure B.7. Section 8, SH 240 And Section 25 SL 473, Wichita Falls.

Figure B.8. Section 9, SH 6, Waco.

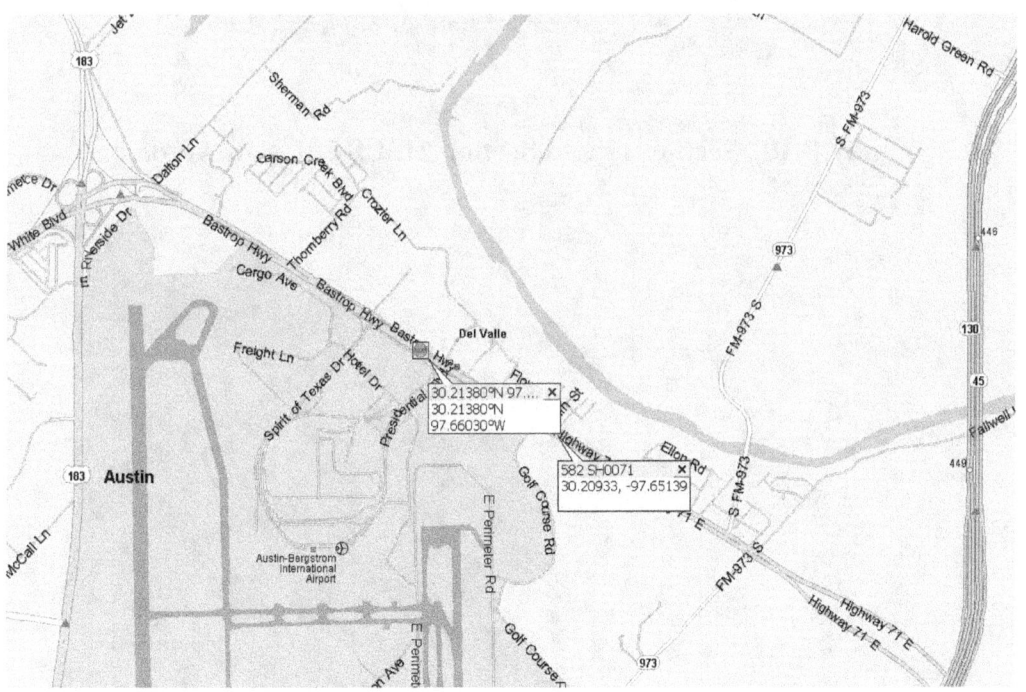

Figure B.9. Section 10, SH 71, Austin.

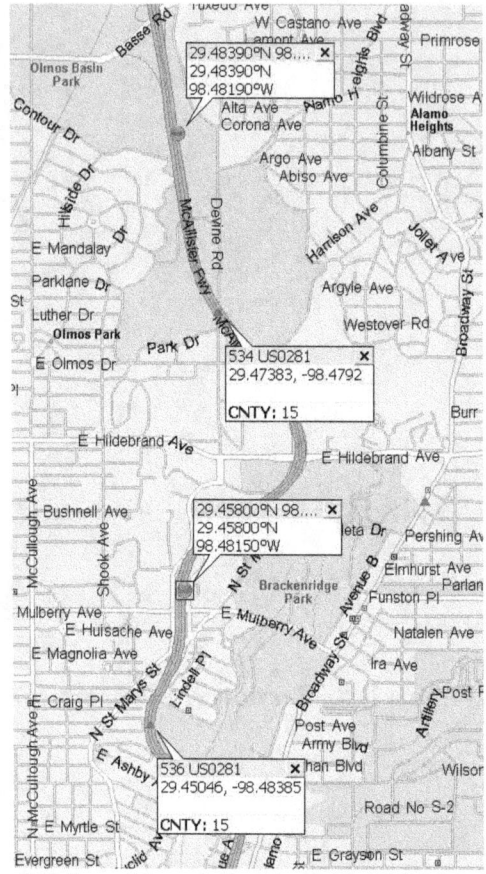

Figure B.10. Section 11 and Section 21, US 281, San Antonio.

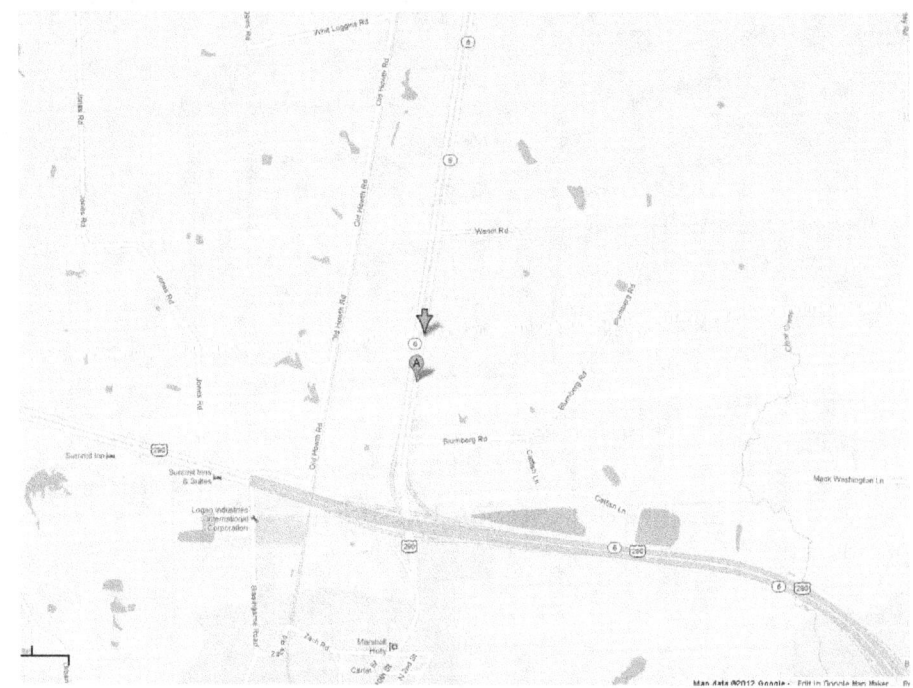

Figure B.11. Section 112, SH 6, Houston.

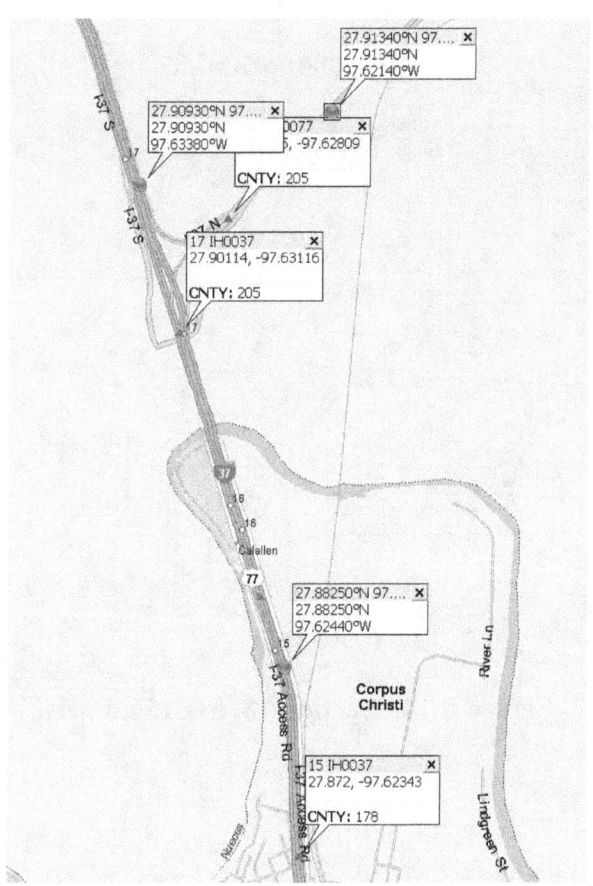

Figure B.12. Sections 14 and 15, IH 37, and Section 16 US 77, Corpus Christi.

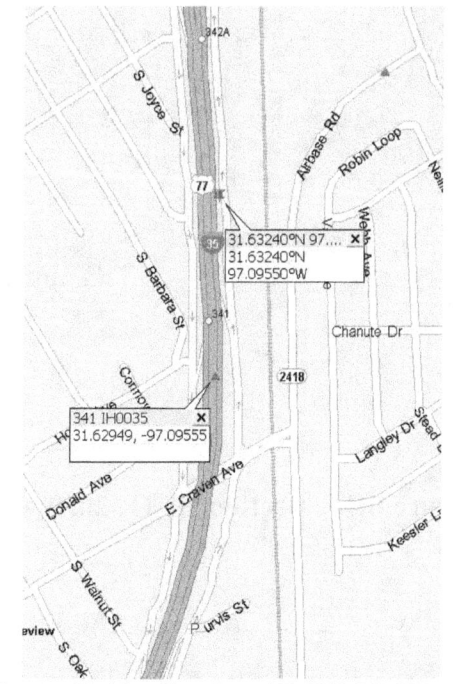

Figure B.13. Section 17, IH 35, Waco.

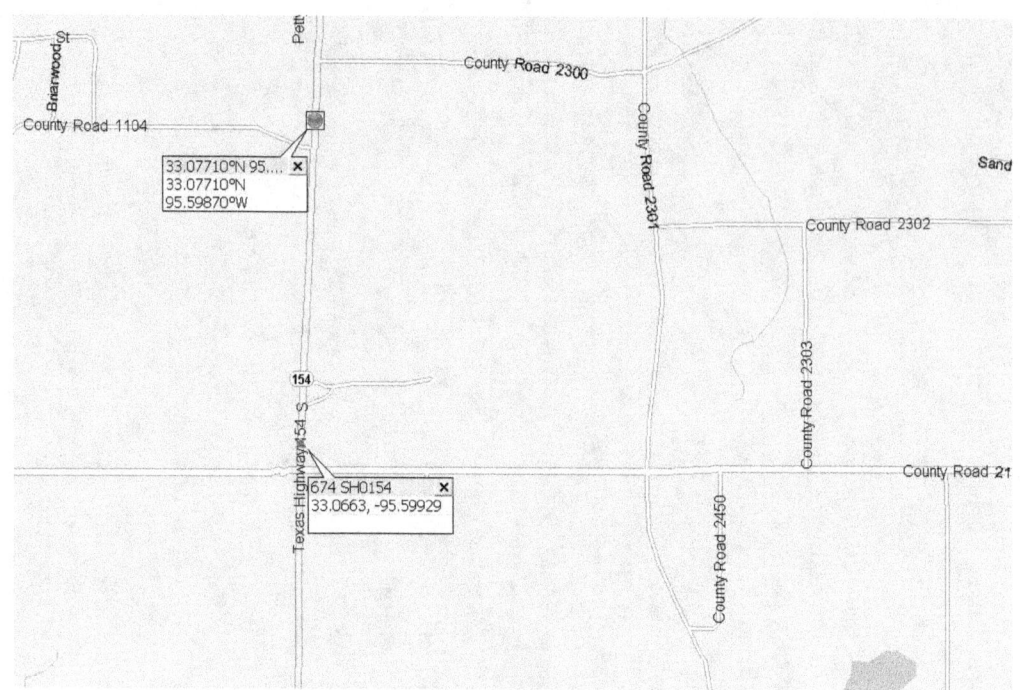

Figure B.14. Section 18, SH 154, Paris.

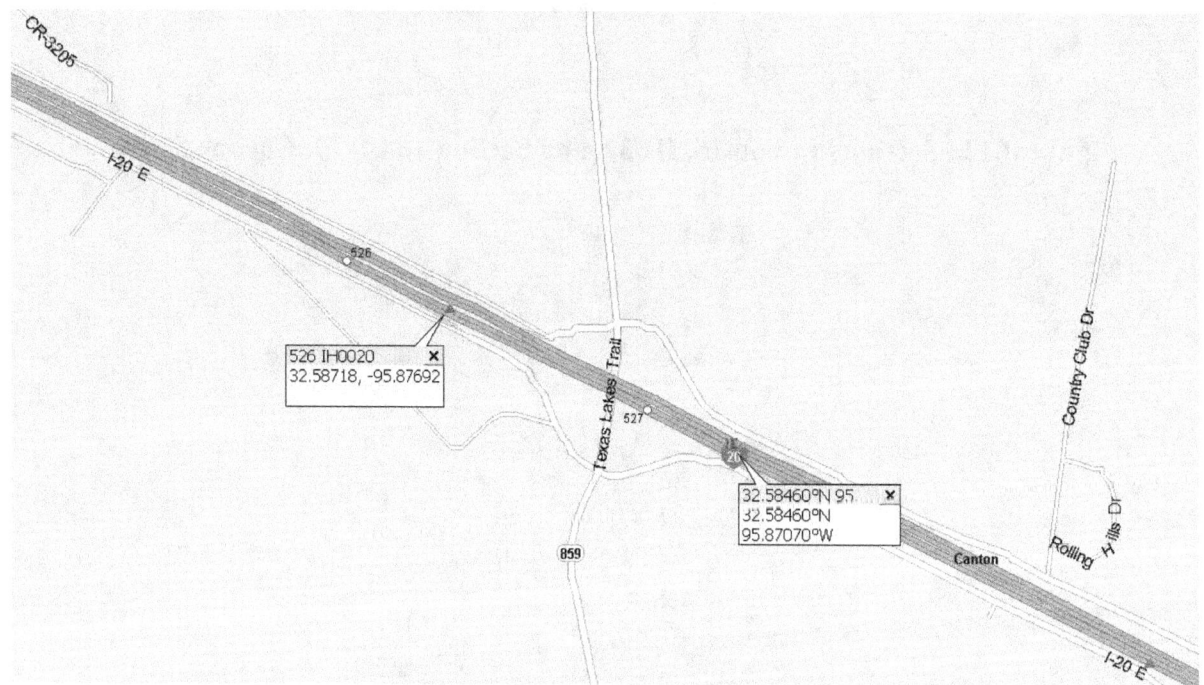

Figure B.15. Section 19, IH 20, Tyler.

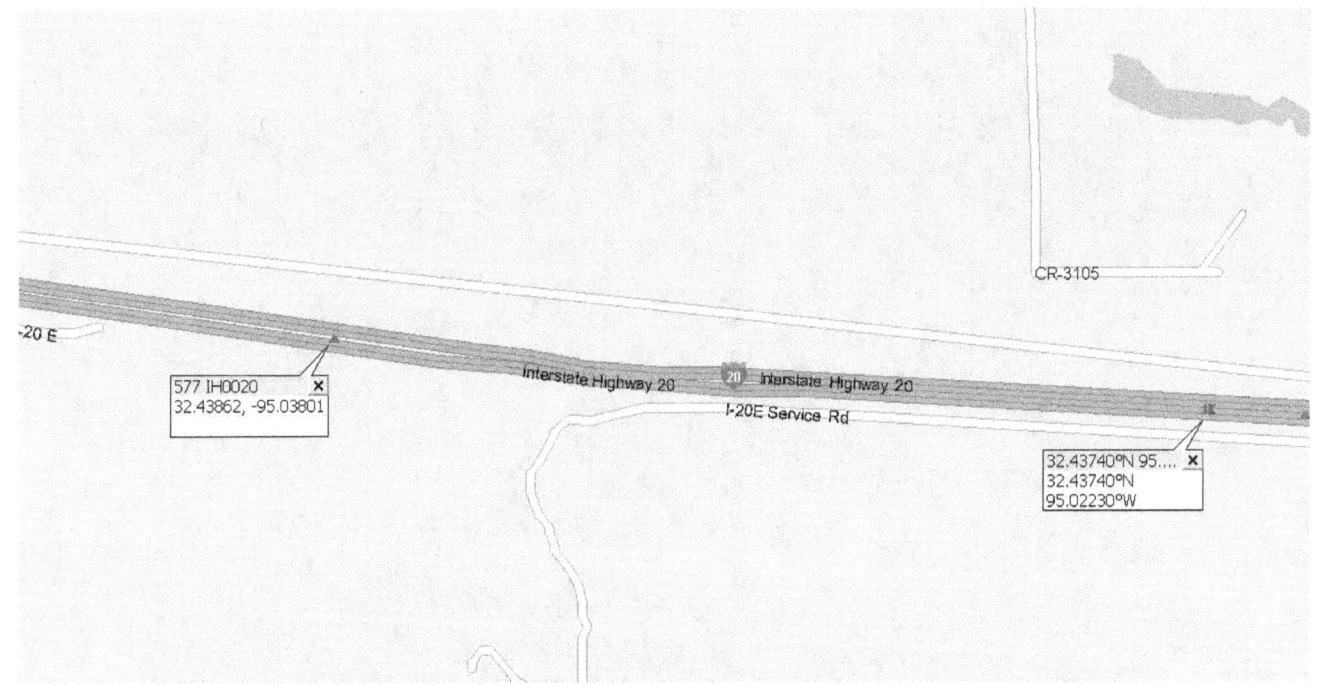

Figure B.16. Section 20, IH 20, Tyler.

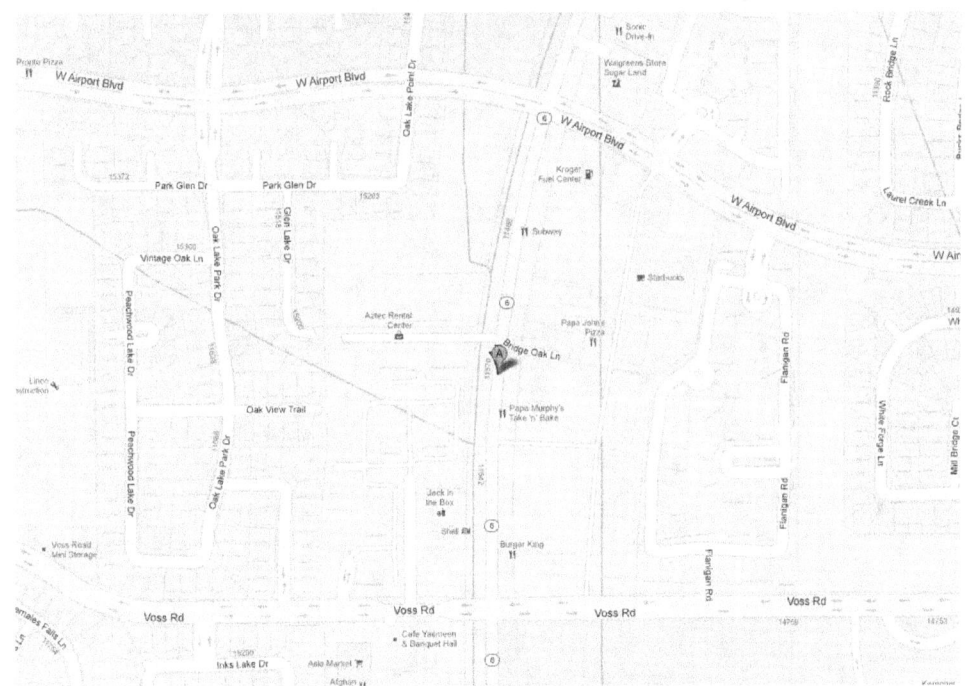

Figure B.17. Section 22, SH 6, Houston.

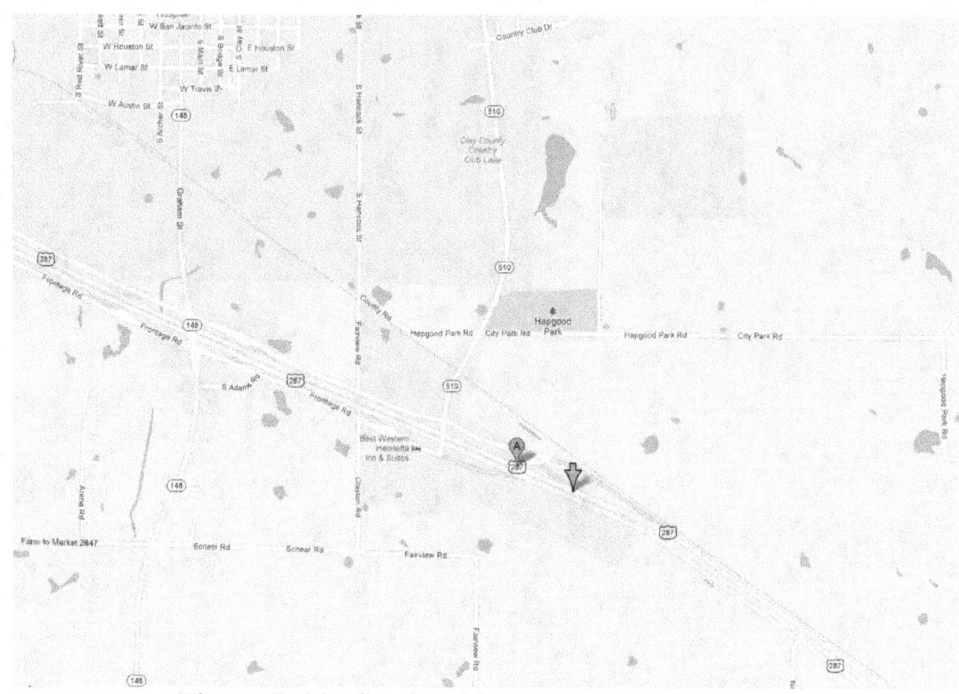

Figure B.18. Section 23, US 287, Wichita Falls.

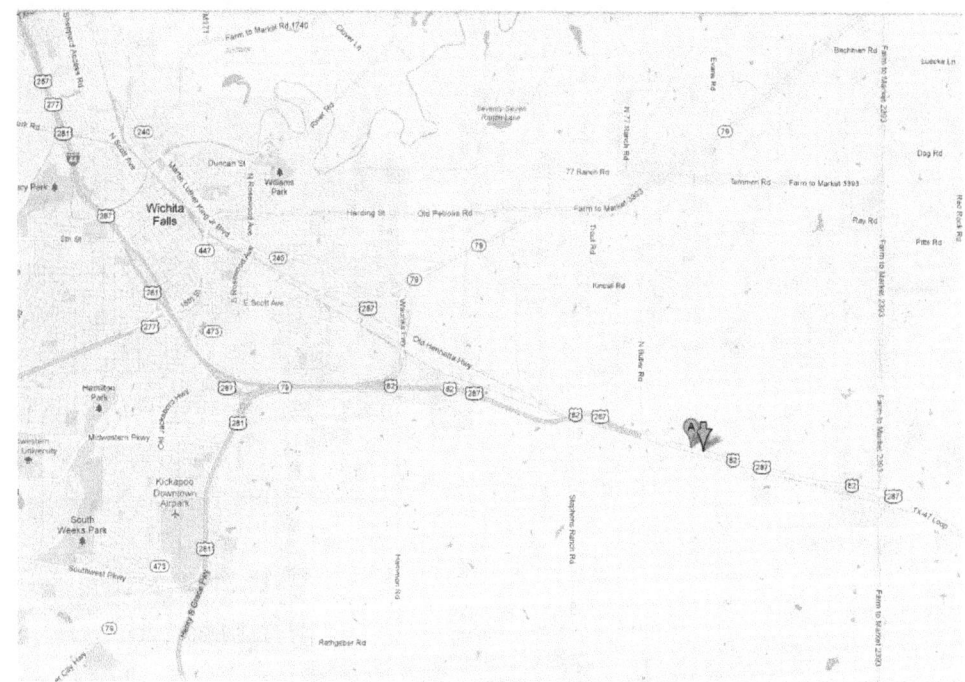

Figure B.19. Section 24, US 82, Wichita Falls.

Figure B.20. Section 26, US 281, Pharr.

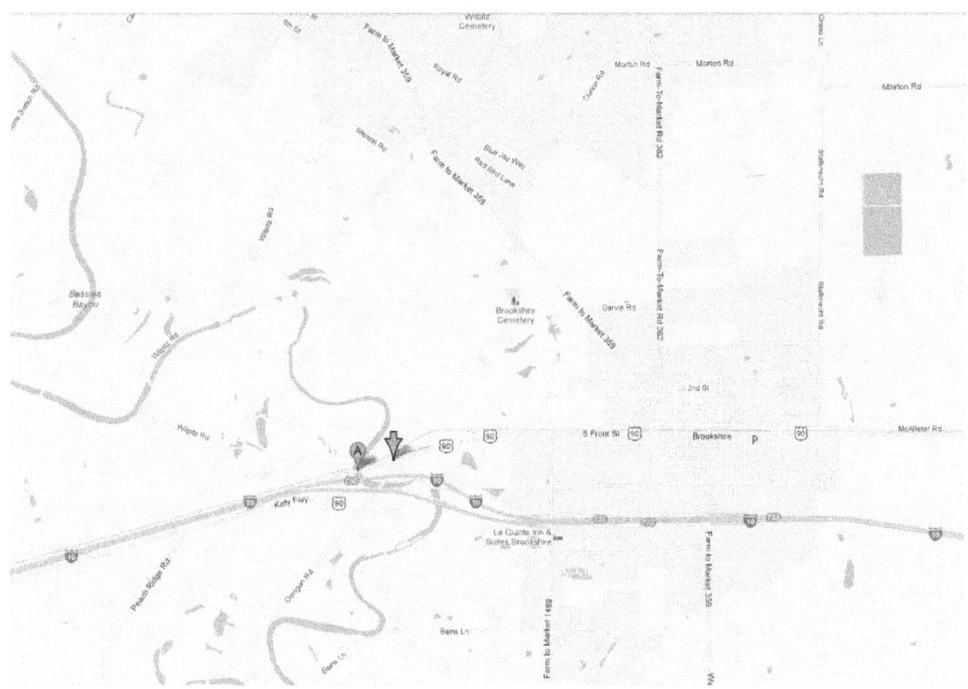
Figure B.21. Section 27, US 90, Houston.

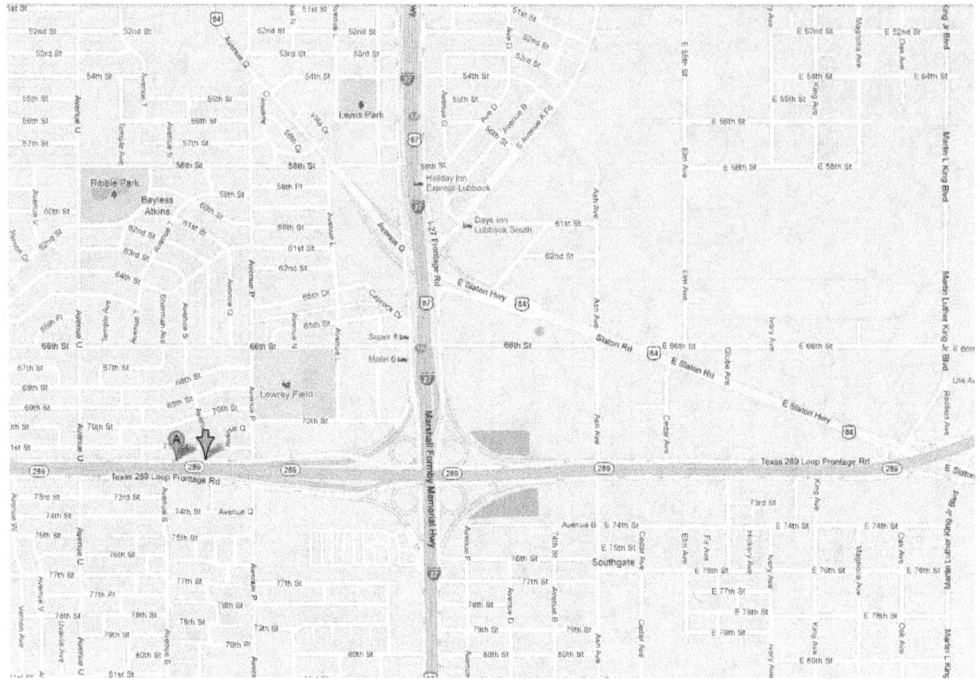

Figure B.22. Section 28, SL 289, Lubbock.

APPENDIX C:
PAVEMENT SECTIONS USED FOR NOISE MEASUREMENTS

Appendix C describes the pavement sections that were evaluated by means of the OBSI method. Approximately 20 sections were measured on several occasions, which included PFC pavements, dense-graded HMA reference pavements, and one of the rotating pavement sections.

The sections' limits utilized for noise measurement differ slightly from those utilized for the other tests that are part of this project. The reason for the difference is that the other tests conducted as part of the project (i.e., WFV, texture, friction, and skid) required a small stretch of pavement to perform the measurements and were done on only one direction of the pavement section. Noise sections, on the other hand, require about 134 m (440-ft) long subsections, and normally, two or three subsections were used on both traveling directions. Thus, noise sections include a large stretch of pavement, in order to accommodate all the subsections required to test the variability within each pavement. In all cases, the lane measured was only the outside lane, while the test vehicle was driven in the center of the lane; the noise fixture was attached to the rear passenger side tire. Therefore, the right wheel path was measured in these tests.

Some sections' characteristics changed over time and those changes caused modifications to the measurements, such as segments being overlaid, repairs and lane closures, and even some sections had to be dropped throughout the course of the research study because noise testing was no longer possible, and those had to be substituted with different pavements.

A description of the PFC sections is presented in the next paragraphs. The description includes location, subsections, and some photographs of the highway views and pavements.

SECTION 1 AND SECTION 13, US 59, YOAKUM

This section of highway in the Yoakum District, near Hillje, featured two pavement types, as each direction had a different type of surface. Therefore, two pavement sections for this study were evaluated in the same highway. The northbound (NB) section corresponds to Section 1 of this project. The southbound (SB) pavement is a dense-graded HMA that was used as a reference pavement (i.e., Section 13); Figure C.1 has a map showing the location. Views of the sections on US 59 are shown in Figures C.2–C.4.

Figure C.1. Sections 1 (PFC) and 13 (REF) on US 59, Yoakum.

Figure C.2. PFC Pavement Section on US 59 (Section 1).

Figure C.3. HMA Pavement Section on US 59 (Section 13).

Figure C.4. Northbound View of US 59.

Table C.1 lists the coordinates of both sections.

Table C.1. Location Coordinates for Section 1.

Subsection	Latitude	Longitude
1-NB1	N29.1352°	W96.3661°
1-NB2	N29.1388°	W96.3603°

SECTION 2, SH 288, HOUSTON

This AR-PFC section was located in the Houston District, near the towns of Manvel and Iowa Colony, south of the City of Houston. Its location is indicated in the map presented in Figure C.5. Four subsections were identified for noise testing, two in the NB direction (2-3 and 2-4), and two in the SB direction (2-1 and 2-2).

Figure C.5. Section 2 on SH 288, Houston.

Figure C.6. View of SH 288, Section 2, Southbound.

Figure C.7. View of SH 288, Section 2, Subsection 2-4, Northbound.

Figure C.8. Detailed View of the PFC Surface on SH 288, Section 2, Northbound.

Table C.2 lists the coordinates of this section.

Table C.2. Location Coordinates for Section 2.

Subsection	Latitude	Longitude
2-SB-1	N29.4086°	W95.4274°
2-SB-2	N29.4022°	W95.4273°
2-NB-3	N29.4022°	W95.4267°
2-NB-4	N29.4088°	W95.4268°

SECTION 3, US 290, AUSTIN

This AR-PFC section was located east of the City of Austin, near Paige, in Bastrop Count (see Figure C.9). Six subsections were located for OBSI testing: three in the eastbound direction (3-1, 3-2, and 3-3), and three in the westbound direction (3-4, 3-5, and 3-6). Pictures of the highway and pavement details are shown in Figures C.10–C.12.

Figure C.9. Section 3 on US 290, Austin.

Figure C.10. View of US 290, Section 3, Subsection 3-1, Eastbound.

Figure C.11. View of US 290, Section 3, Subsection 3-3, Eastbound.

Figure C.12. Detailed View of the PFC Surface on US 290, Section 3, Subsection 3-3, EB.

Table C.3 shows the coordinates for Section 3.

Table C.3. Location Coordinates for Section 3.

Subsection	Latitude	Longitude
3-1	N30.2076°	W97.0925°
3-2	N30.2011°	W97.0792°
3-3	N30.1993°	W97.0756°
3-4	N30.1995°	W97.0756°
3-5	N30.2012°	W97.0791°
3-6	N30.2077°	W97.0924°

SECTION 4, IH 30, PARIS

This PFC, designated as Section 4, is located in the Paris District, east of Sulphur Springs (see Figure 29). Six subsections were marked for noise tests: 4-1, 4-2, and 4-3 in the westbound (WB) direction, and 4-4, 4-5, and 4-6 in the eastbound (EB) direction. Figures C.14–C.17 present views of the highway and the pavement.

Figure C.13. Section 4 on IH 30, Paris.

Figure C.14. View of IH-30, Section 4, Subsection 4-1, Westbound.

Figure C.15. Detailed View of the PFC Surface on IH-30, Section 4, Subsection 4-1, WB.

Figure C.16. View of IH-30, Section 4, Subsection 4-5, Eastbound.

Figure C.17. Detailed View of the PFC Surface on IH-30, Section 4, Subsection 4-6, EB.

Table C.4 lists the coordinates for Section 4.

Table C.4. Location Coordinates for Section 4.

Subsection	Latitude	Longitude
4-1	N33.1479°	W95.4512°
4-2	N33.1469°	W95.4682°
4-3	N33.1469°	W95.4786°
4-4	N33.1466°	W95.4769°
4-5	N33.1467°	W95.4682°
4-6	N33.1512°	W95.4381°

SECTION 5, SH 6, BRYAN

Designated as Section 5 in this project, this new PFC was constructed in May 2009 in the Bryan District and became the second newest pavement tested in this study. The newest PFC is Section 20, on IH-20, in Tyler, with a construction date of August 2009. The section is located north of Calvert. This project consists of six subsections: three in the northbound (5-1, 5-3, and 5-4) and three in the southbound (5-5, 5-6, and 5-7) directions (see Figure C.18). Some images taken from the section are shown in Figures C.19–C.21.

Figure C.18. Section 5 on SH 6, Bryan.

Figure C.19. View of SH 6, Section 5, Subsection 5-1, Northbound.

Figure C.20. Detailed View of the PFC Surface on SH 6, Section 5, Subsection 5-1, NB.

Figure C.21. Views of SH 6, Section 5, Southbound.

Table C.5 lists the coordinates for Section 5.

Table C.5. Location Coordinates for Section 5.

Subsection	Latitude	Longitude
5-1	N31.0328°	W96.7092°
5-3	N31.0371°	W96.7096°
5-4	N31.0430°	W96.7101°
5-5	N31.0430°	W96.7106°
5-6	N31.0371°	W96.7101°
5-7	N31.0329°	W96.7098°

SECTION 6, IH 20, ABILENE

This PFC, designated as Section 6 in this project, is located in Abilene, near the intersection with US 83, close to Section 7, which is subsequently described. There are four subsections identified for noise testing: two WB (6-1, and 6-2), and two EB (6-3 and 6-4). As Figure C.22 shows, subsections 6-2 and 6-3 are west of US 83, while 6-1 and 6-4 are east of that highway. Figures C.23 and C.24 show images from Section 6. Table C.6 presents the coordinates for Section 6.

Table C.6. Location Coordinates for Section 6.

Subsection	Latitude	Longitude
6-1	N32.4885°	W99.7553°
6-2	N32.4803°	W99.7793°
6-3	N32.4789°	W99.7830°
6-4	N32.4870°	W99.7579°

Figure C.22. Section 6 on IH 20, Abilene.

Figure C.23. View of IH 20, Section 6, Eastbound.

Figure C.24. Detailed View of the PFC Surface on IH 20, Section 6, Subsection 6-1, WB.

SECTION 7, US 83, ABILENE

Section 7, as well as the preceding section, is located in Abilene. Five subsections were identified as part of this PFC project: three in the NB lanes (7-1, 7-2, and 7-3) and two in the SB lanes (7-4 and 7-5). This pavement section is located in the proximity of IH-20, with one subsection (7-3) north of the interstate, and the remaining four subsections south of it, as Figure C.25 shows. Pictures of the section appear in Figures C.26–C.28.

Figure C.25. Section 7 on US 83, Abilene.

Figure C.26. View of US 83, Section 7, Subsection 7-1, Northbound.

Figure C.27. Detailed View of the PFC Surface on US 83, Section 7, Subsection 7-2, NB.

Figure C.28. View of US 83, Section 7, Subsection 7-4, Southbound.

The Table C.7 presents the coordinates for Section 7.

Table C.7. Location Coordinates for Section 7.

Subsection	Latitude	Longitude
7-1	N32.4560°	W99.7856°
7-2	N32.4697°	W99.7792°
7-3	N32.4938°	W99.7611°
7-4	N32.4808°	W99.7713°
7-5	N32.4726°	W99.7775°

SECTION 9, SH 6, WACO

This PFC on SH 6 just east of the City of Waco, in the Waco District, was constructed in 2005, and it started being measured for noise in 2006, as part of TxDOT Project 0-5185. Figure C.29 shows the location of the pavement section, and the four subsections identified for OBSI tests: WB-1 and WB-2, in the westbound direction, and EB-3 and EB-4, in the eastbound direction. Figures C.30–C.32 show various aspects of this section.

Figure C.29. Section 9 on SH 6, Waco.

Figure C.30. View of SH6, Section 9 on SH 6, Westbound.

Figure C.31. View of SH6, Section 9, Subsection 9.

Figure C.32. Detailed View of the PFC Surface on SH 6, Waco.

The coordinates for Section 9 are presented in Table C.8.

Table C.8. Location Coordinates for Section 9.

Subsection	Latitude	Longitude
WB-1	N31.5410°	W97.0481°
WB-2	N31.5434°	W97.0558°
EB-3	N31.5448°	W97.0621°
EB-4	N31.5424°	W97.0539°

SECTION 11 AND SECTION 21, US 281, SAN ANTONIO

Two adjacent PFC sections on US 281 in San Antonio were measured in this study as well as in prior TxDOT Project 0-5185. Both of them are AR-PFCs and are among the quietest pavements in the state. Because these sections have been studied since 2006, it can be said that they have remained quieter than other pavements throughout the years, even though this urban highway carries a high amount of traffic.

Section 11, the northernmost of the two, was constructed in 2005, and it is the quietest PFC that has been measured since 2006. The section is comprised of six subsections: 3T, 4T, and 5T in the SB lanes, and 12T, 13T, and 14T in the NB lanes. TxDOT identified and assigned the numbers for these subsections in conjunction with the adjacent section. Figure C.33 shows these locations in the proximity of Basse Rd. Figures C.34 and C.35 show pictures from this section.

Figure C.33. Section 11 on US 281, San Antonio.

Figure C.34. View of US 281, Section 11, Subsection 12T, Northbound.

Figure C.35. Detailed View of the PFC Surface on US 281, Section 11, Northbound.

Section 21 was adjacent to Section 11, located just south of it. Constructed in 2006, this pavement was included in this project as one of the rotating pavement sections. Based on the measurements acquired as part of prior TxDOT Project 0-5185, this PFC was very quiet. From a testing standpoint, the advantage of having side-to-side sections is that both were tested in the same field trip. However, these sections had a geometry disadvantage. As Figure C.36 shows, there are several horizontal curves that, coupled with the presence of numerous exit and entrance ramps, made it one of the most difficult stretches of road to measure at test speeds. Seven subsections comprise Section 21: 6T, 7T, 8T, and 17T, in the SB direction, and 9T, 10T, and 11T, in the NB direction. Views from Section 21 are presented in C.37 and C.38.

Figure C.36. Section 21 on US 291, San Antonio.

Figure C.37. US 281, Section 21, Subsection 8T, Southbound.

Figure C.38. Detailed View of the PFC Surface on US 281, Section 21, Subsection 17T, SB.

The coordinates for Sections 11 and 21 are presented in Table C.9.

Table C.9. Location Coordinates for Sections 11 and 21.

Section	Subsection	Latitude	Longitude
11	3T	N29.5787°	W29.0354°
	4T	N29.3152°	W28.9943°
	5T	N28.9014°	W28.9180°
	12T	N29.0330°	W28.9164°
	13T	N29.1533°	W28.9390°
	14T	N29.5392°	W29.0099°
21	6T	N29.4680°	W98.4746°
	7T	N29.4614°	W98.4791°
	8T	N29.4604°	W98.4808°
	17T	N29.4531°	W98.4828°
	9T	N29.4580°	W98.4815°
	10T	N29.4638°	W98.4747°
	11T	N29.4655°	W98.4736°

SECTION 112, SH 6, HOUSTON

This PFC pavement, designated as Section 112 in this project, was added to this study in 2010 as a substitute for Section 12, which was located on IH-10 in Yoakum District, but was overlaid after the 2009 measurements. The pavement is located in Hempstead, just north of the intersection with US 290, in the Houston District (see Figure C.39). Six subsections—three in each direction—comprise this project with regard to noise measurements. The NB subsections are labeled 112-1, 112-2, and 112-3, and the SB subsections are referred to as 112-4, 112-5, and 112-6. These are shown on the map in Figure C.39. Pictures of the section appear in Figures C.40–C.42.

Figure C.39. Section 112 on SH 6, Houston.

Figure C.40. View of SH 6, Section 112, Northbound.

Figure C.41. View of SH 6, Section 112, Southbound.

Figure C.42. Detailed View of the PFC Surface on SH 6, Section 112, Southbound.

Table C.10 lists the coordinates for Section 112.

Table C.10. Location Coordinates for Section 112.

Subsection	Latitude	Longitude
112-1	N30.1229°	W96.0744°
112-2	N30.1250°	W96.0740°
112-3	N30.1272°	W96.0736°
112-4	N30.1327°	W96.0730°
112-5	N30.1296°	W96.0736°
112-6	N30.1255°	W96.0743°

SECTION 14 AND SECTION 15, IH 37, CORPUS CHRISTI

Two adjacent PFC sections on IH-37 in Corpus Christi District were measured as part of this project. The northernmost PFC is in San Patricio County; its limits span the Nueces River Bridge to the Atascosa County line. This section was identified as Section 14, and its binder is PG. The southernmost PFC is in Nueces County. Its limits extend from downtown Corpus Christi at US 281 to north of the Nueces River Bridge. This section is an AR-PFC, and it was designated as Section 15 in this project. Both were constructed in 2004. Besides the measurements in this project, CTR started investigating their noise properties in 2006, as part of the aforementioned TxDOT Project 0-5185. When first marked on the pavement, these sections were identified as CC2 (Section 14) and CC1 (Section 15). The subsections identified for noise testing are near the towns of Odem and Calallen. Five of them were marked: two in the SB direction, and three in the NB lanes. Figure C.43 shows the map with the various subsection. Figures C.44–C.47 show general and detailed views of the surface of the pavement for these two sections.

Figure C.43. Sections 14 (CC2) and 15 (CC1) on IH 37, Corpus Christi.

Figure C.44. Views of IH-37, Section 14 (CC2), Northbound.

Figure C.45. Detailed View of the PFC Surface on IH-37, Section 14, Corpus Christi.

Figure C.46. Views of IH-37, Section 15 (CC1), Northbound.

Figure C.47. Detailed View of the AR-PFC Surface on IH-37, Section 15, Corpus Christi.

The coordinates of the sections are listed in Table C.11.

Table C.11. Location Coordinates for Sections 14 and 15.

Section	Subsection	Latitude	Longitude
14	CC2NB1	N27.9115°	W97.6346°
	CC2NB2	N27.9201°	W97.6383°
15	CC1NB2	N27.8824°	W97.6243°
	CC1SB1	N27.8889°	W97.6270°
	CC1SB2	N27.8709°	W97.6235°

SECTION 17, IH 35, WACO

Constructed in 2003, this section crosses through downtown Waco, near Craven Ave. It became the oldest PFC in this study, after the previous oldest section, the original Section 12 of this Project, located on IH-10 in Yoakum District, constructed in 2001, was overlaid around 2009. The IH-35 pavement is designated as Section 17 in this project. It consists of four subsections: two in the NB lanes and two in the SB lanes. Figure C.48 shows the project location.

On a trip to this section around June 2008, the research team noticed that two of the original PFC subsections had been overlaid with new PFC pavement. The new overlay was placed on the southernmost subsections, designated as 17-NB1 and 17-SB2. During the last visit to this section in summer 2012, there was ongoing construction on the old PFC (the northernmost subsections). Some aspects of the PFC subsections are shown in Figures C.49–C.51.

Figure C.48. Section 17 on IH 35, Waco.

Figure C.49. View of IH 35, Section 17, Subsection NB-1.

Figure C.50. View of IH 35, Section 17, Northbound (Old PFC).

Figure C.51. View of IH 35, Section 17, Subsection SB-2 (New PFC).

The coordinates for the IH-35 Waco subsections are listed in Table C.12.

Table C.12. Location Coordinates for Section 17.

Subsection	Latitude	Longitude
17-NB1	N31.6177°	W97.0999°
17-NB2	N31.6325°	W97.0955°
17-SB1	N31.6336°	W97.0960°
17-SB2	N31.6206°	W97.0989°

SECTION 19, IH 20, TYLER

This PFC in the Tyler District was identified as Section 19 in this study. There are three EB subsections (19-1, 19-2, and 19-3), and three WB subsections (19-4, 19-5, and 19-6) in this project (Figure C.52). Figures C.53 and C.54 show some views of the IH-20 section. Table C.13 shows the coordinates for Section 19.

Figure C.52. Section 19, IH 20, Tyler.

Figure C.53. View of IH 20, Section 19, Subsection 19-4, Westbound.

Figure C.54. Detailed View of the PFC Surface on IH 20, Section 19, Subsection 19-5, WB.

Table C.13. Location Coordinates for Section 19.

Subsection	Latitude	Longitude
19-1	N32.5933°	W95.8922°
19-2	N32.5871°	W95.8768°
19-3	N32.5808°	W65.8614°
19-4	N32.5811°	W95.8613°
19-5	N32.5882°	W95.8787°
19-6	N32.5936°	W95.8920°

SECTION 20, IH 20, TYLER

Section 20, on IH-20 in Tyler District, is the newest PFC pavement analyzed in this study. The PFC was constructed in August 2009. Figure C.55 shows the map with the project location. Figures C.56–C.58 show pictures of the section. Table C.14 presents the coordinates for Section 20.

Figure C.55. Section 20 on IH 20, Tyler.

Figure C.56. View of IH 20, Section 20, Subsection 20-1, Eastbound.

Figure C.57. Detailed View of the PFC Surface on IH 20, Section 20, Subsection 20-3, WB.

Figure C.58. View of IH 20, Section 20, Subsection 20-3, Westbound.

Table C.14. Location Coordinates for Section 20.

Subsection	Latitude	Longitude
20-1	N32.4374°	W95.0223°
20-2	N32.4363°	W94.9887°
20-3	N32.4367°	W94.9905°
20-4	N32.4377°	W95.0197°

APPENDIX D:
OBSI TEST RESULTS FOR INDIVIDUAL PAVEMENT SECTIONS

Appendix D gives a detailed description of the OBSI test results obtained for each of the permanent and one rotating pavement section. The overall noise levels and frequency spectra are described.

SECTION 1 AND SECTION 13, US 59, YOAKUM

This section of highway in the Yoakum District, near Hillje, featured two different pavement sections: Section 1 corresponded to the northbound (NB) direction, while the southbound (SB) pavement corresponded to Section 13 (dense-graded HMA). The overall noise levels for Section 1 and frequency spectra for both pavement sections are shown in Figures D.1 and D.2, respectively.

An important difference in the shape of the spectral curves can be observed between the PFC (Section 1) and the dense-graded HMA pavement (Section 13), even though the overall levels for both sections are very similar. The PFC was expected to be quieter than its dense-graded counterpart was. Contrary to what was anticipated, the overall nose levels were evenly matched throughout the years. From a visual standpoint, the NB PFC appeared to be in better condition than the SB reference section.

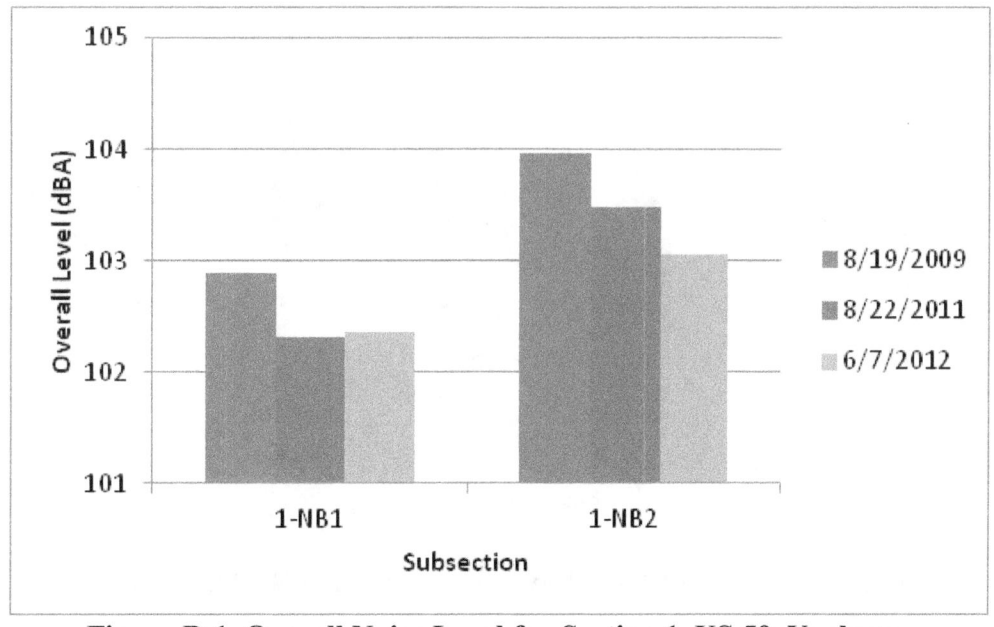

Figure D.1. Overall Noise Level for Section 1, US 59, Yoakum.

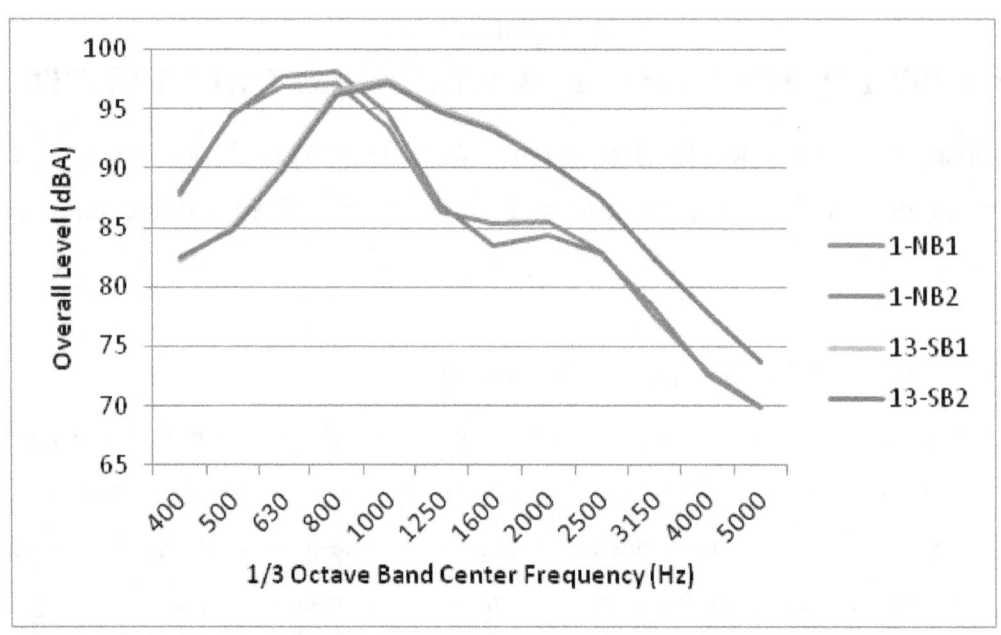

Figure D.2. Frequency Spectra for Sections 1 and 13 Measured 6/7/2012.

SECTION 2, SH 288, HOUSTON

The measurements for 2012 in this section yielded erroneous results due to a faulty microphone. Unfortunately, those measurements had to be discarded. The overall levels for Section 2 are shown in Figure 21 shows the overall levels for Section 2, and Figure 22 presents the frequency spectra for the subsections corresponding to the 2011.

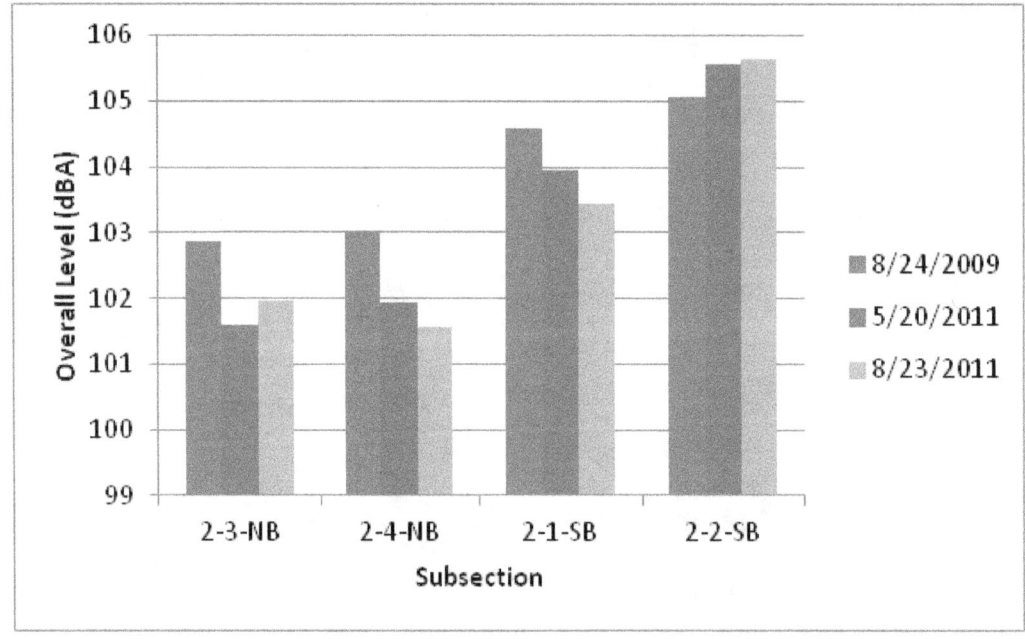

Figure D.3. Overall Noise Level for Section 2, SH 288, Houston.

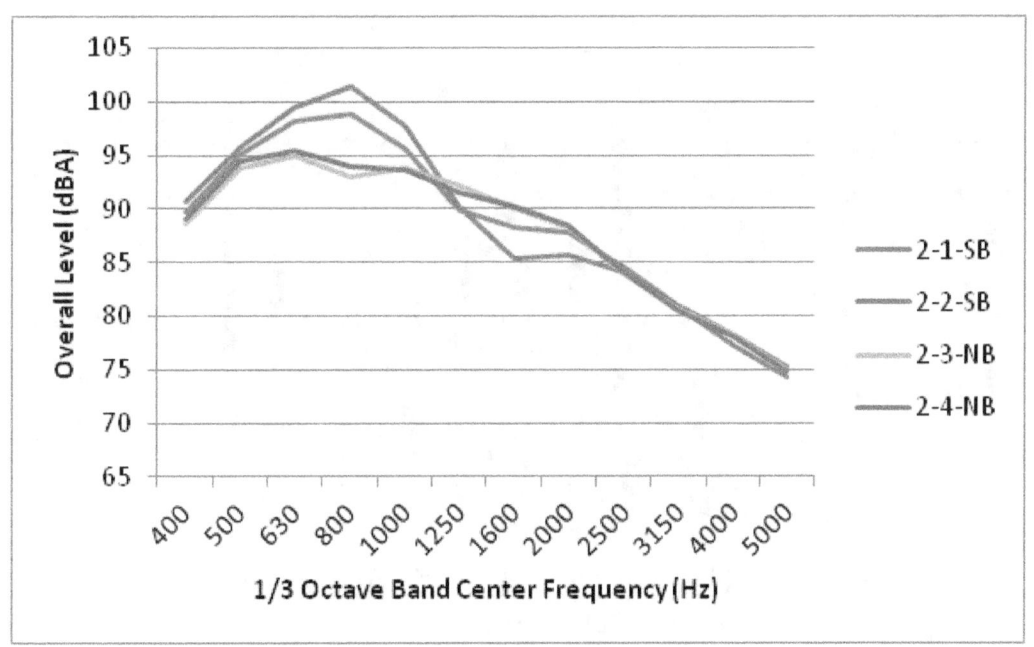

Figure D.4. Frequency Spectra for Section 2 Measured 8/23/2011.

This section presented substantial differences in overall levels between SB and NB subsections, and thus ranks among the highest in variability within subsections for the study. The acoustical differences between traveling directions were even more evident in the frequency spectra curves. The NB subsections were quieter and had flatter spectra, whereas the SB subsections presented a pronounced peak at 800 Hz. The peak is common at these frequencies for PFCs, but the magnitude at that frequency is high, resulting in the higher overall noise levels. This was the loudest pavement among the AR-PFCs included in the project.

SECTION 3, US 290, AUSTIN

The measured noise levels for Section 3 throughout the years are presented in Figure D.5, and the most recent frequency graph is shown in Figure D.6. When the 2010 noise measurements are disregarded, the average noise level stayed consistent throughout the study, around 101 dBA. A good reason to disregard the 2010 result, not only for this section, is that the 2010 set of tests were conducted with the oldest tire, the AWP, as mentioned in Chapter 3. The average overall level for this section matches the average AR-PFC level for this project (101.4 dBA).

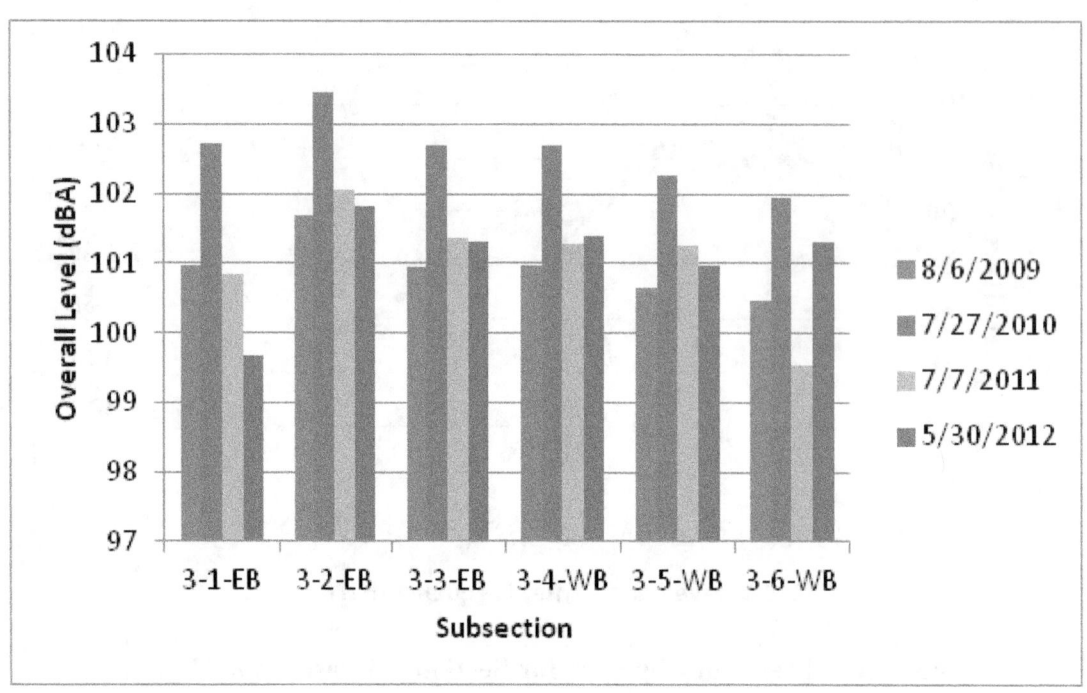

Figure D.5. Overall Noise Level for Section 3, US 290, Austin.

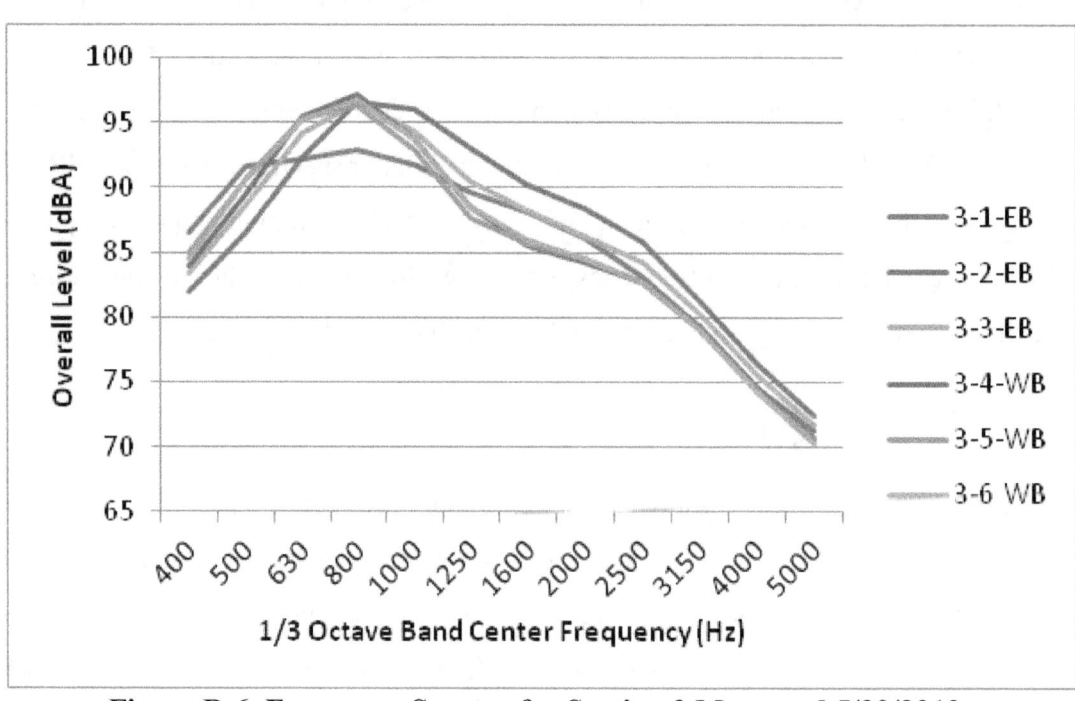

Figure D.6. Frequency Spectra for Section 3 Measured 5/30/2012.

SECTION 4, IH 30, PARIS

The overall noise levels for the IH-30 PFC are shown in Figure D.7, and the most recent frequency spectra chart, from 2012, is displayed in Figure D.8.

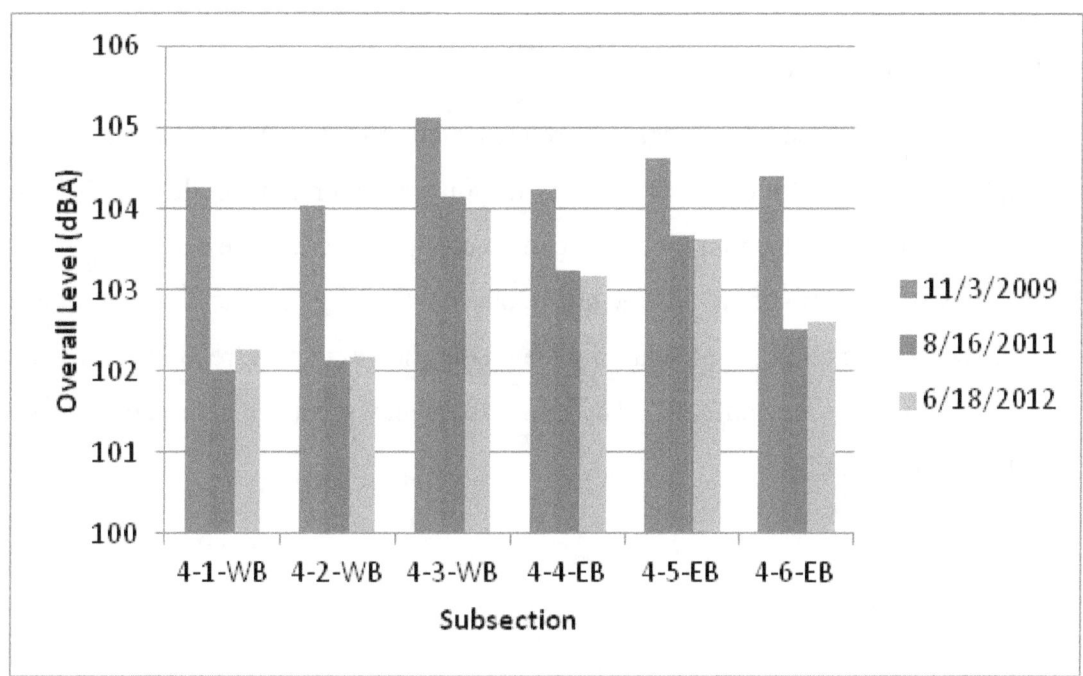

Figure D.7. Overall Noise Levels for Section 4, IH 30, Paris.

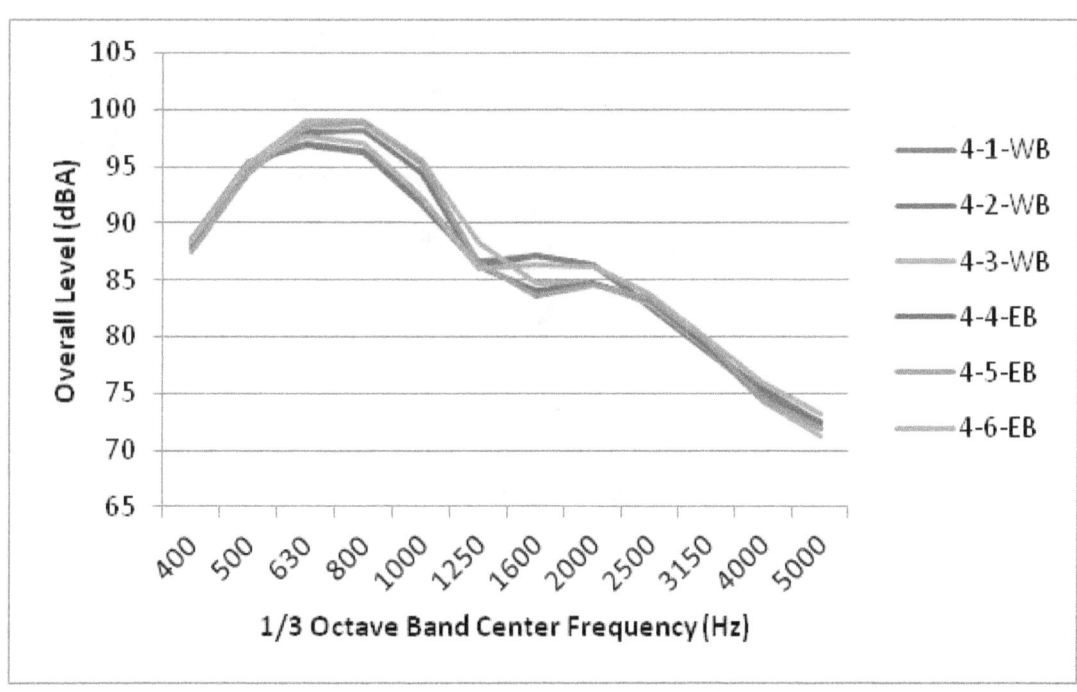

Figure D.8. Frequency Spectra for Section 4 Measured 6/18/2012.

This section was one of the loudest PFCs in this study, perhaps due to the fact that it is a TBPFC. Other studies have shown that the thickness of the pavement has an influence in noise absorption, with thicker layers being quieter. Also, the noise levels in measurement years 2011 and 2012 were lower, but this may be due to the influence of the ages of the test tires.

SECTION 5, SH 6, BRYAN

The OBSI measurements on Section 5 are presented in Figure D.9, and the most recent frequency spectra are shown in Figure D.10. This section was among the loudest PFCs in this project, despite being one of the more recently constructed pavements. The surface looked in good shape, shows no deterioration, and the traffic observed during noise testing was sparse. The first time the section was tested, the SB lanes were louder; however, this trend was reversed in the last two sets of tests (2011 and 2012). The recent frequency spectra show relatively uniform shapes for the subsections.

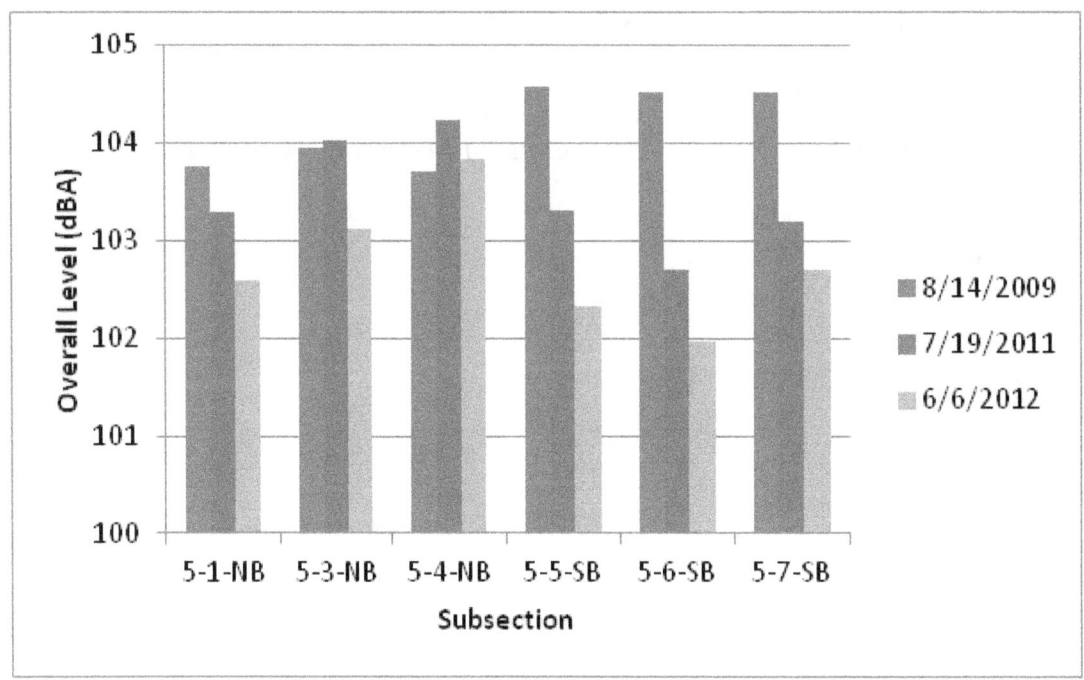

Figure D.9. Overall Noise Level for Section 5, SH 6, Bryan.

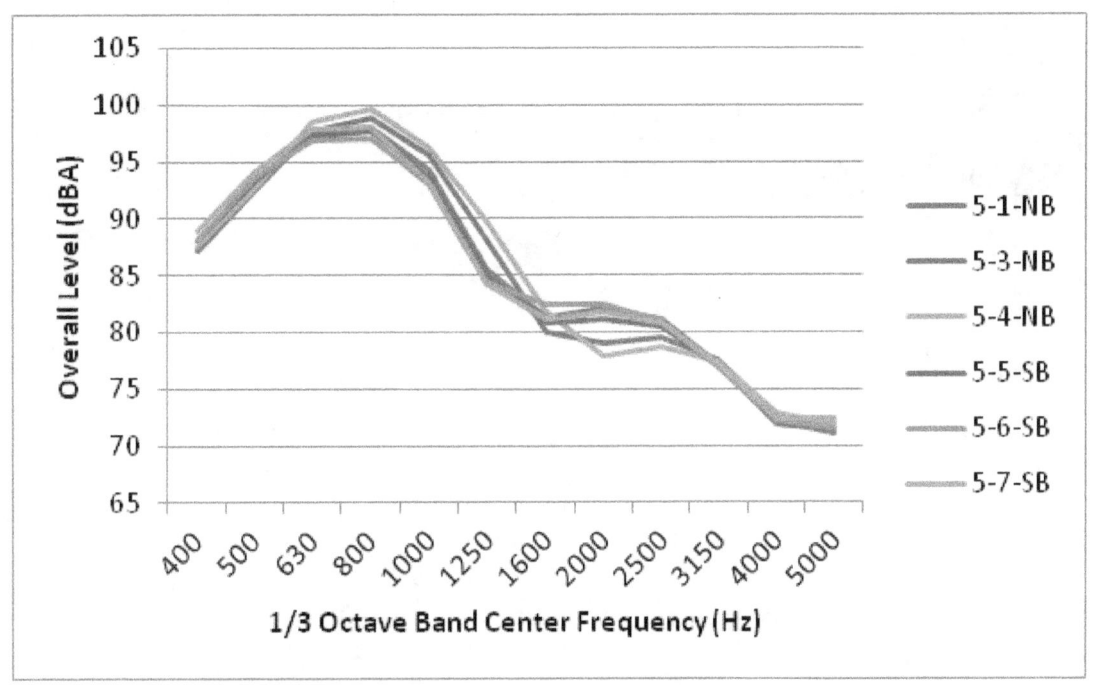

Figure D.10. Frequency Spectra for Section 5 Measured 6/6/2012.

SECTION 6, IH 20, ABILENE

Measurements on IH-20 in Abilene are summarized in Figures D.11 and D.12.

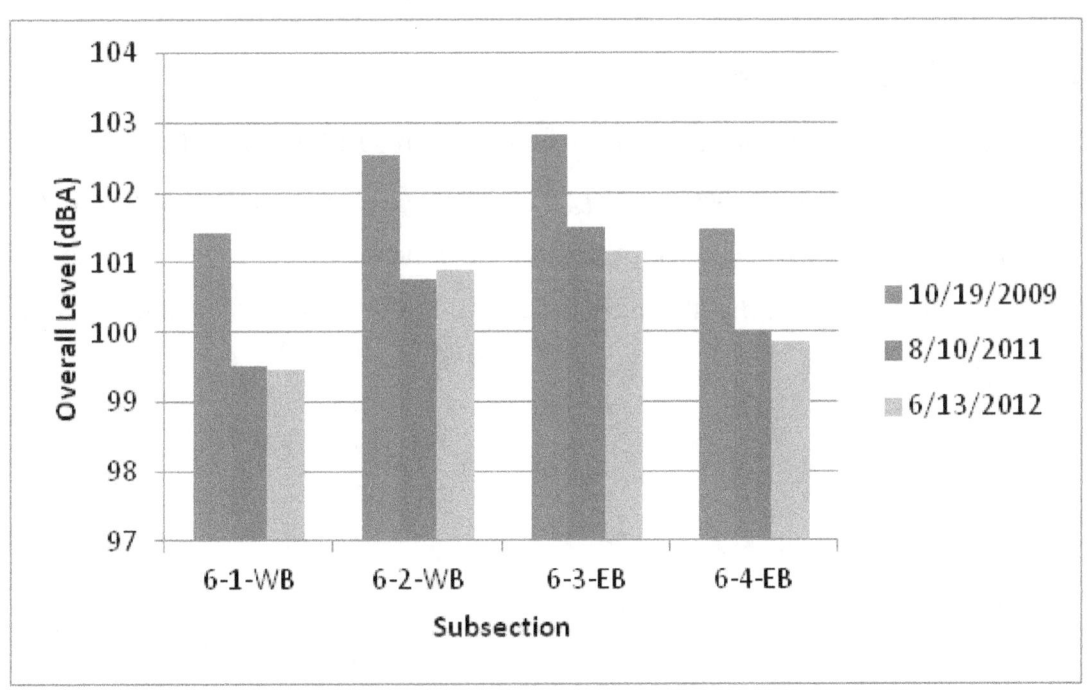

Figure D.11. Overall Noise Level For Section 6, IH 20, Abilene.

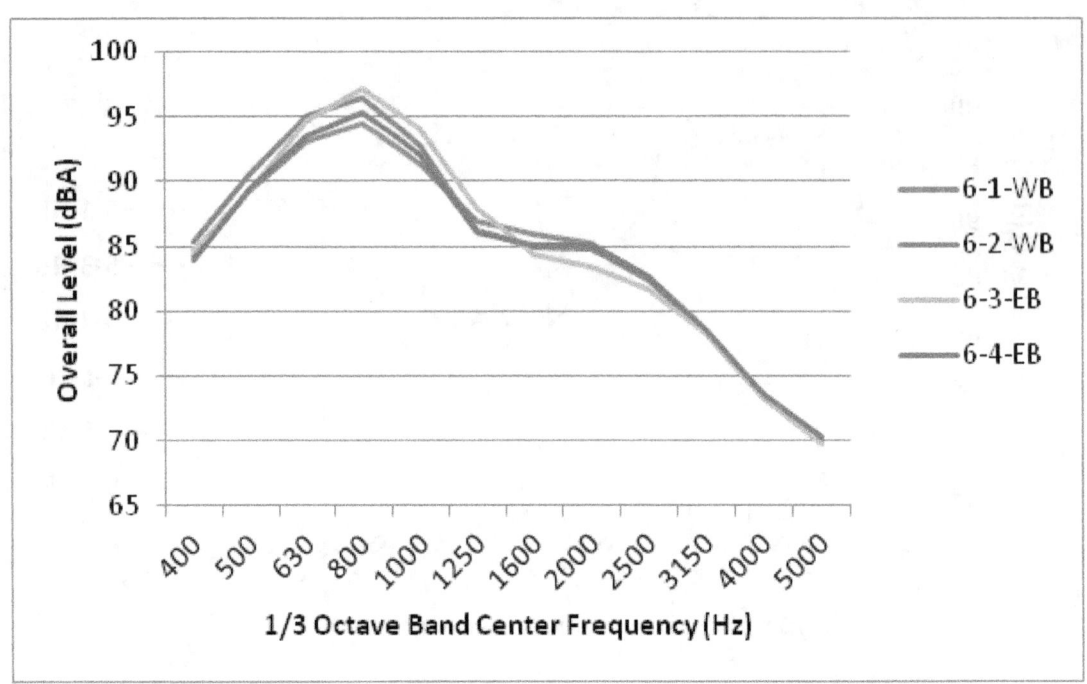

Figure D.12. Frequency Spectra For Section 6 Measured 6/13/2012.

Section 6 in Abilene can be classified as the quietest PG-PFC in this study and is among the quietest pavements. Its average noise level over time is 100.9 dBA. Like several other cases in the project, Section 6 appears to be quieter in measurement years 2011 and 2012.

SECTION 7, US 83, ABILENE

OBSI overall levels for Section 7 are shown in Figure D.13, and the 2012 frequency spectra for the subsections are presented in Figure D.14. Noise levels on this pavement were consistent throughout the years. These levels remained in the middle range with respect to the other PFCs included the study, slightly under the overall PFC average, and well under the overall PG-PFC average.

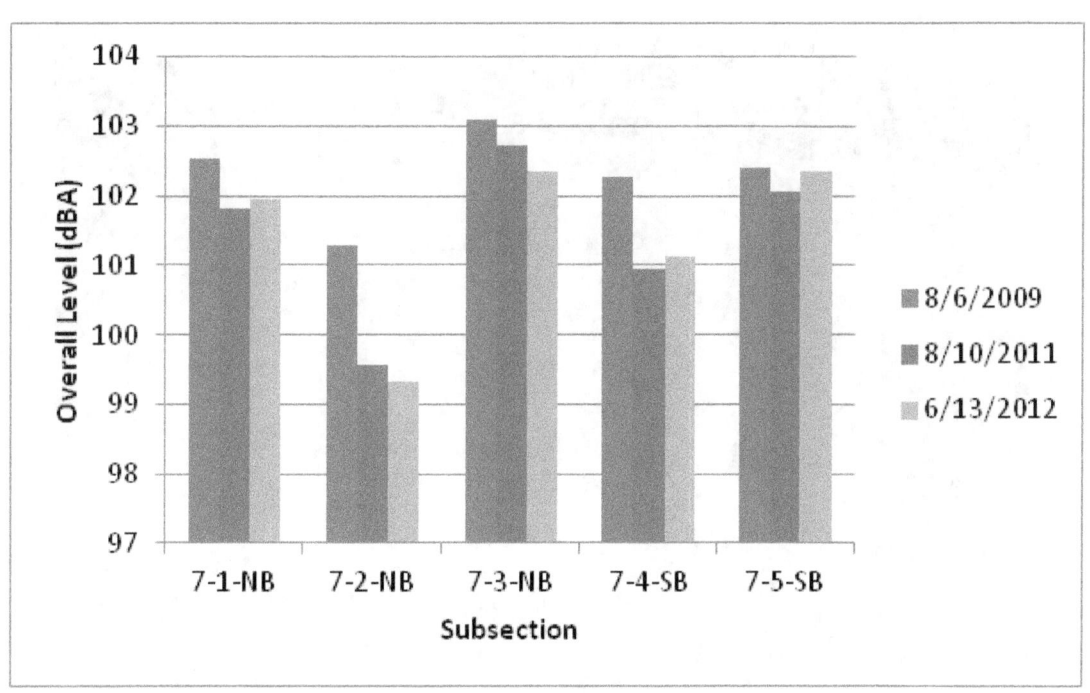

Figure D.13. Overall Noise Level for Section 7, US 83, Abilene.

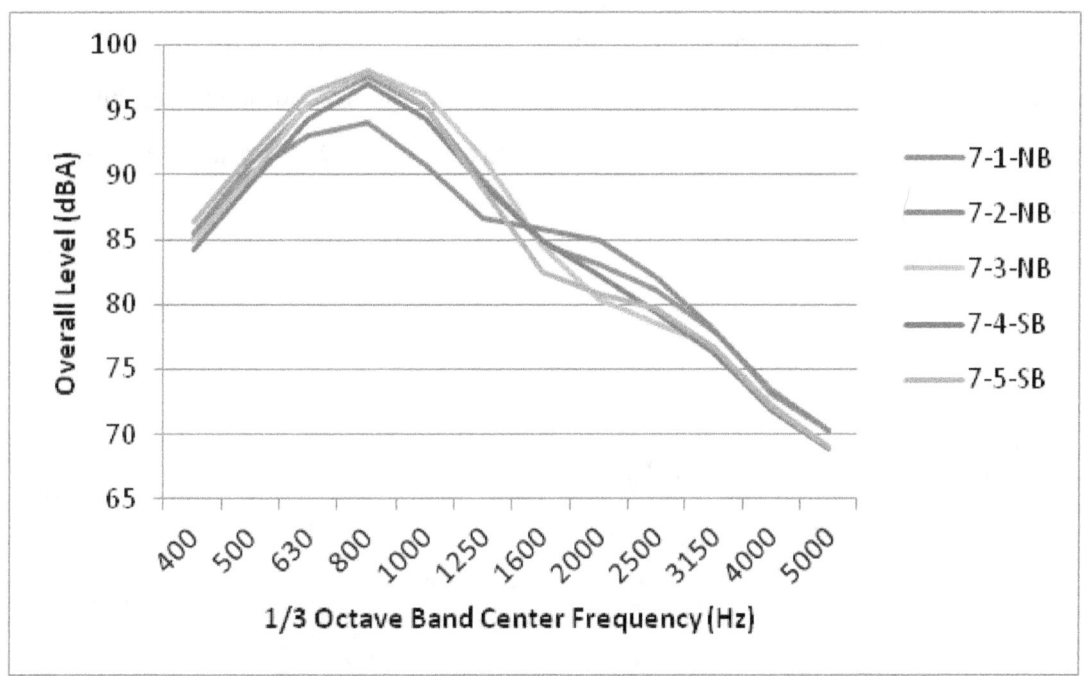

Figure D.14. Frequency Spectra for Section 7 Measured 6/13/2012.

SECTION 9, SH 6, WACO

For the 2012 tests, the subsection WB-1 presented long patches that covered the entire width of the outside lane, as shown in Figure D.15.

Figure D.15. Patch on SH6, Subsection WB-1, Waco.

The presence of such repairs had an impact on the measurements for that particular subsection as shown in Figure D.16, which displays the historic overall levels for the SH 6 Waco subsections, including tests from TxDOT Project 0-5185. Figure D.17 presents the frequency chart for 2012.

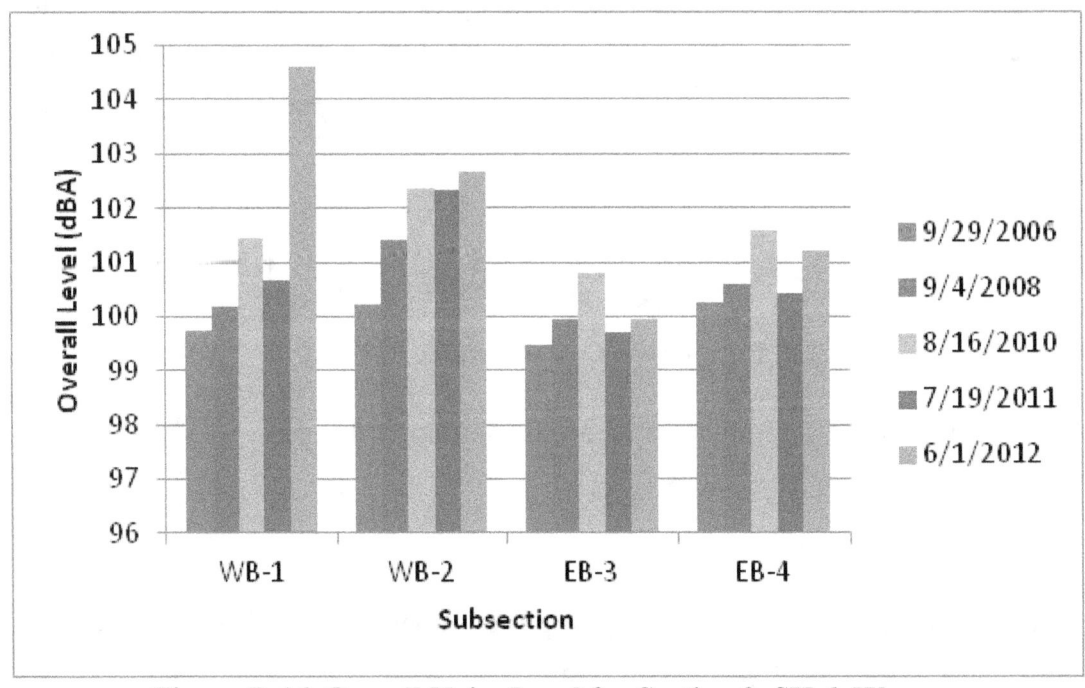

Figure D.16. Overall Noise Level for Section 9, SH 6, Waco.

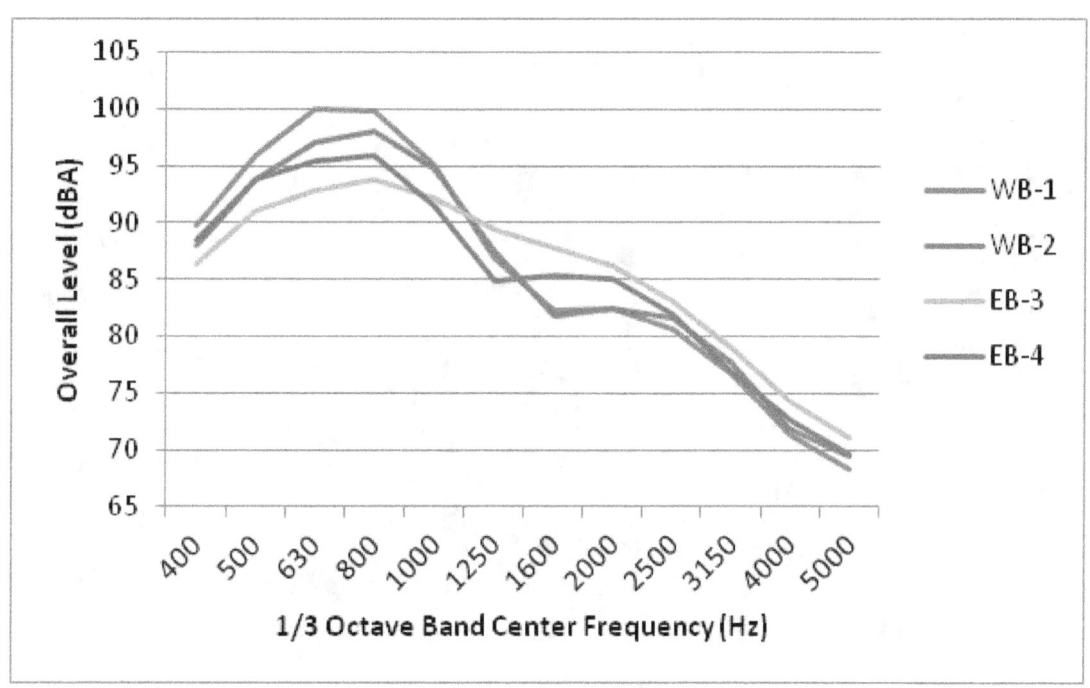

Figure D.17. Frequency Spectra for Section 9 Measured 6/1/2012.

This section was the quietest PG-PFC in this study before the more recent measurements in 2012. Two factors caused an increase in the measured noise levels: 1) severe raveling on this section, possibly due to binder aging or the PFC pavement reaching the end of its service life (constructed in 2005), and 2) the subsequent repairs that unfolded to restore the quality of the riding surface. Also, the variability within subsections went up as a result of the patching and repairs. The latest noise measurements ranked this pavement section near the average level for PFCs in this project.

SECTION 11 AND SECTION 21, US 281, SAN ANTONIO

Overall measurements for Sections 11 and 21 are shown in Figures D.18 and D.20, respectively, and their frequency spectra for 2012 are shown in Figures D.19 and D.21, respectively. These values include tests performed under TxDOT Project 0-5185. Both of these sections are among the quietest in this study and the entire state. The overall frequency spectrum of Section 11 is that of a typical PFC, as opposed to other types of pavements. Even though there is a peak at around 800 Hz, the magnitude is low compared to other pavements. The overall frequency spectrum for Section 21 has an unusual shape with peaks at the 500 and 1000 Hz one-third octave frequency bands, but the magnitude of those are low.

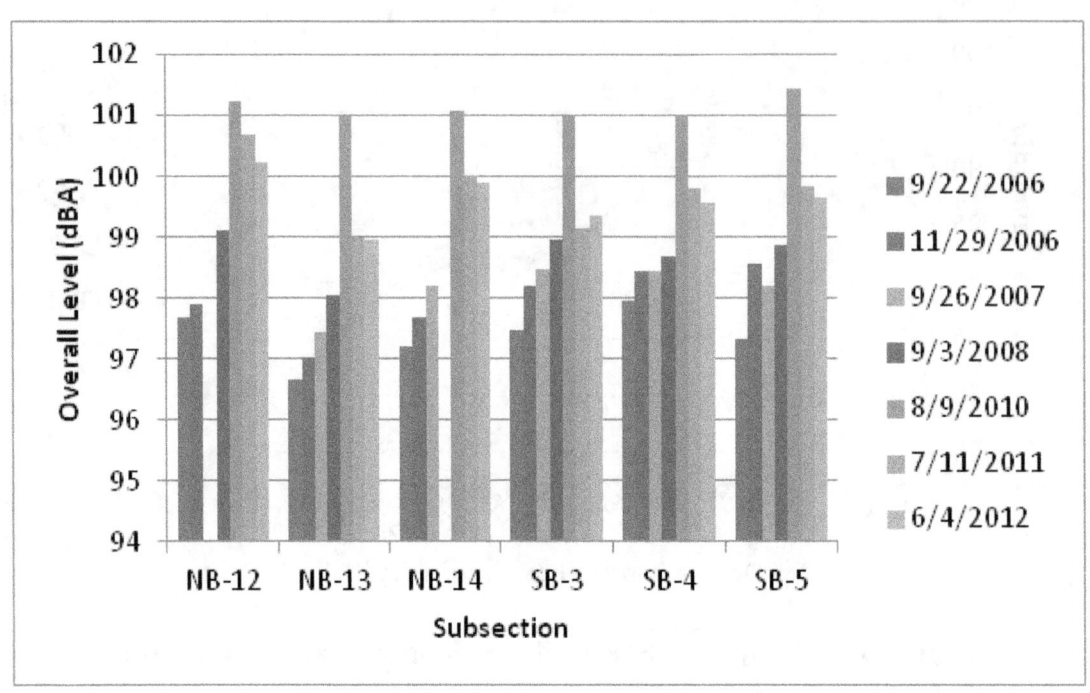

Figure D.18. Overall Noise Level for Section 11, US 281, San Antonio.

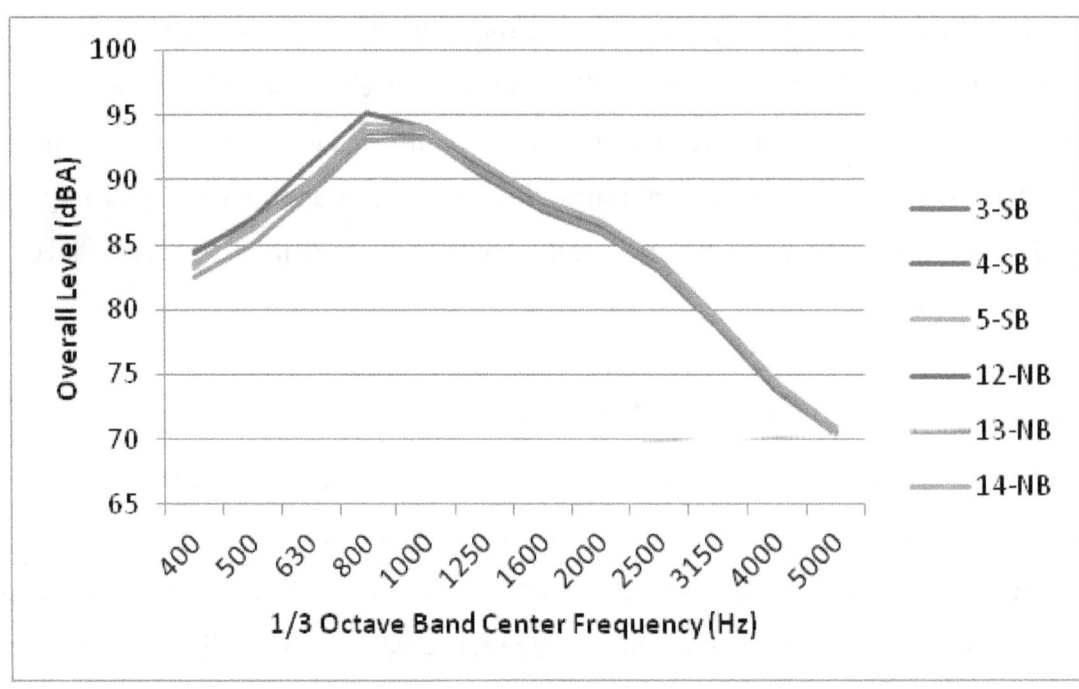

Figure D.19. Frequency Spectra for Section 11 Measured 6/4/2012.

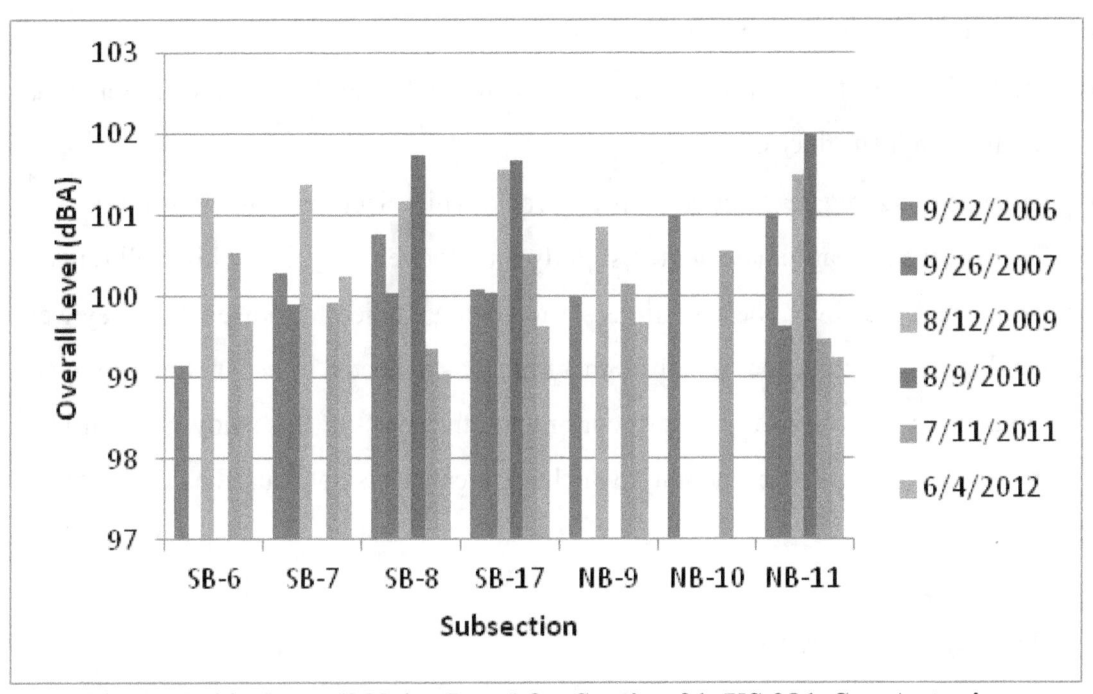

Figure D.20. Overall Noise Level for Section 21, US 281, San Antonio.

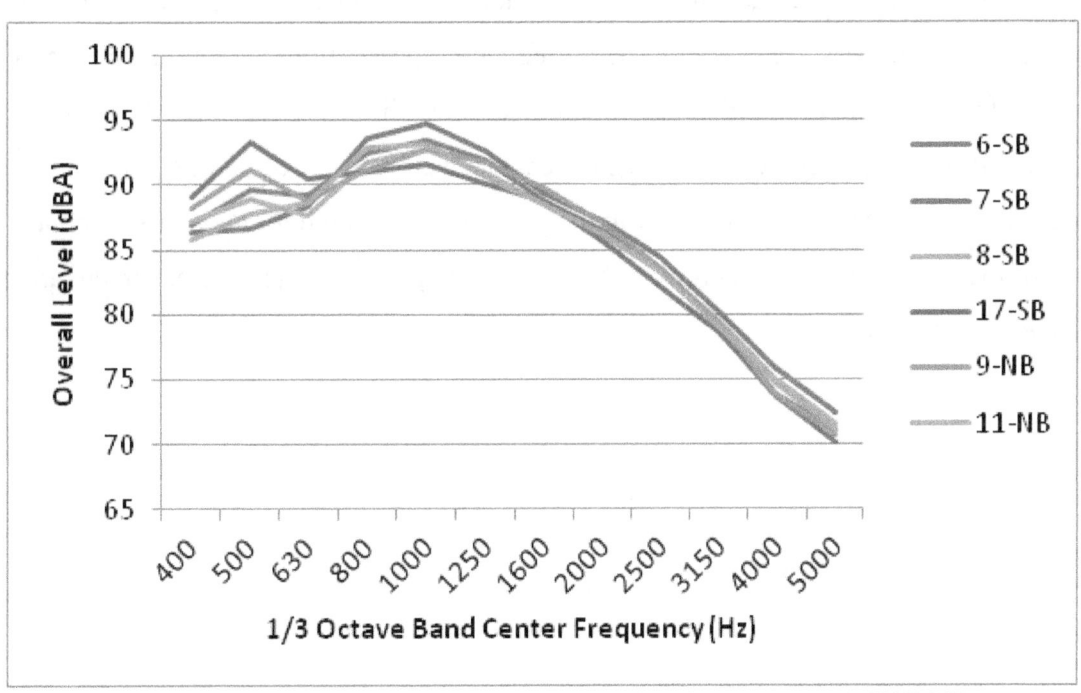

Figure D.21. Frequency Spectra for Section 21 Measured 6/4/2012.

The main differences in both sections are:

- The thickness of the overlays (Section 11 is 50 mm [2 inch] thick and Section 21 is 38 mm [1.5 inch] thick).
- The type of aggregate (Section 11 is trap rock, while Section 21 is sandstone).

The sections appear to have gotten slightly louder over the years, while still remaining among the quietest. However, these sections started being studied for noise when they were almost brand new, so a noise increase from that time was expected to occur, since compaction and clogging took place over the years. According to other parts of this study, Section 11 has been found to have very poor drainability, which apparently has not caused a decline of its acoustic performance.

SECTION 112, SH 6, HOUSTON

Figure D.22 presents the results of the measurements over time for Section 112. The 2012 measurements had to be discarded due to a microphone error, which is the same problem that affected Section 2 in measurement year 2012. Frequency spectra for this section, corresponding to the 2011 tests, are shown in Figure D.23. This section is slightly louder than the average for PG-PFCs in this study. Only two sets of measurements were conducted on this pavement because the section was incorporated late in the study as a substitute for another section and because of the mishap with the microphones in the last round of testing. The pavement appeared to be in good condition.

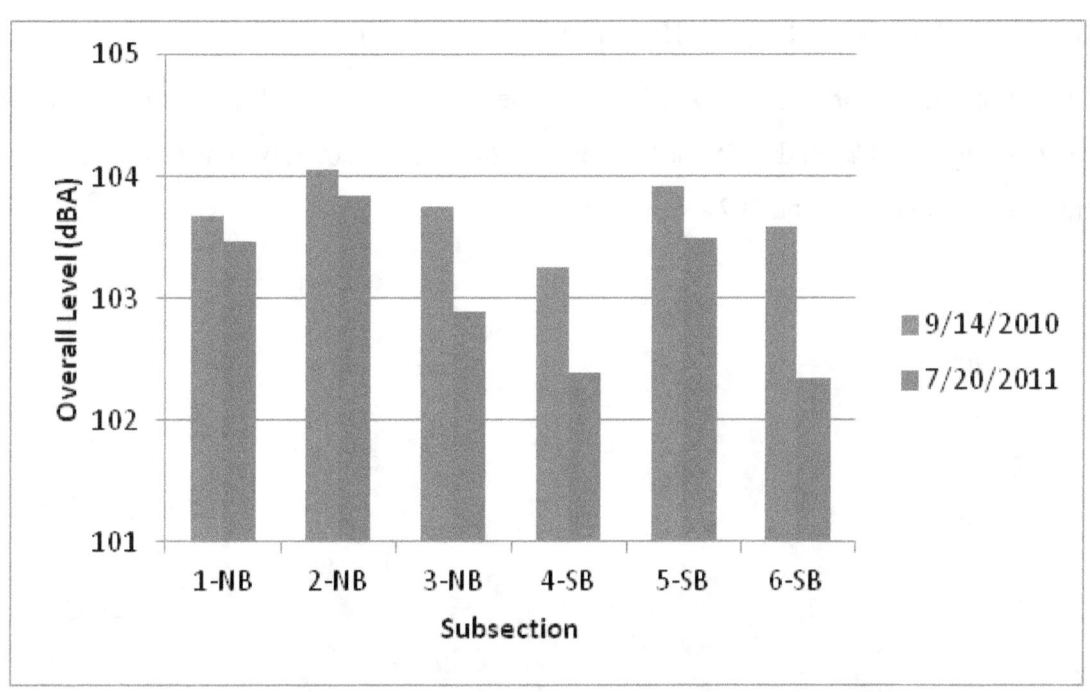

Figure D.22. Overall Noise Level for Section 112, SH 6, Houston.

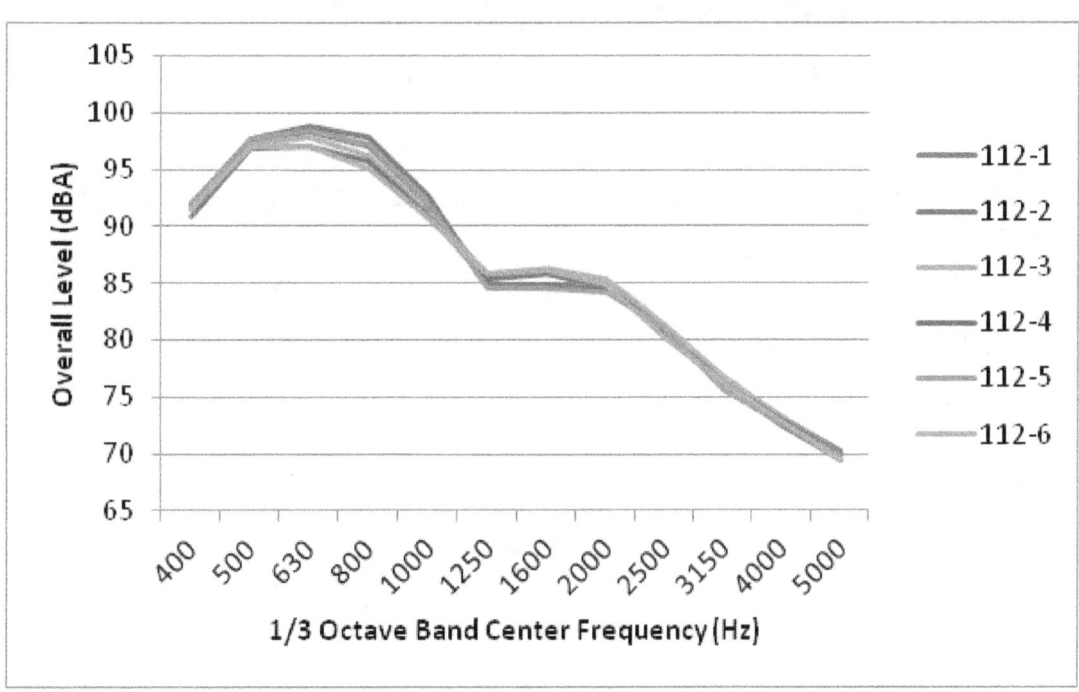

Figure D.23. Frequency Spectra for Section 112 Measured 7/20/2011.

SECTION 14 AND SECTION 15, IH 37, CORPUS CHRISTI

Measurements for these PFCs, which include the historic noise levels from year 2006, are presented in Figures D.24 and D.26 for Sections 14 and 15, respectively. The frequency spectra are shown in Figures D.25 and D.27, respectively.

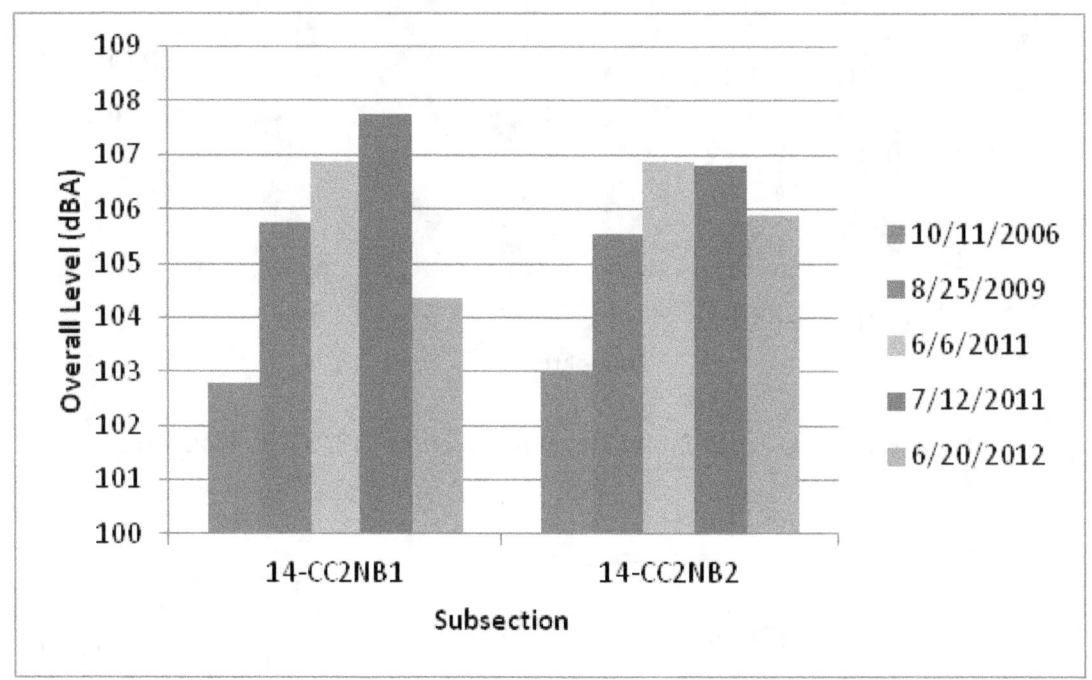

Figure D.24. Overall Noise Level for Section 14, IH-37, Corpus Christi.

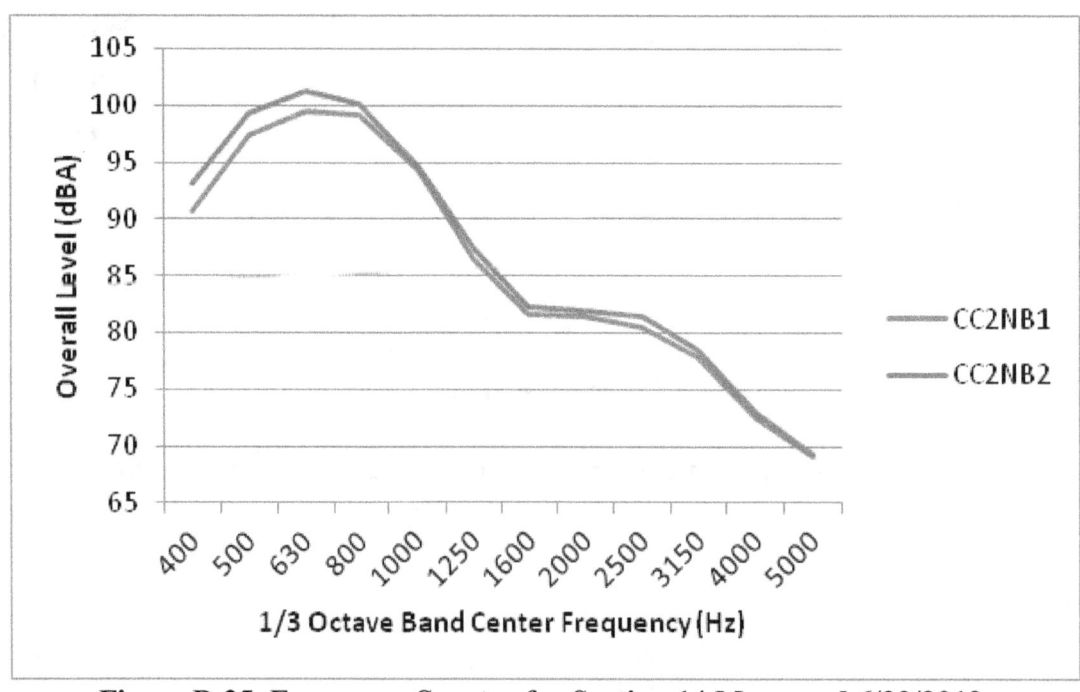

Figure D.25. Frequency Spectra for Section 14 Measured 6/20/2012.

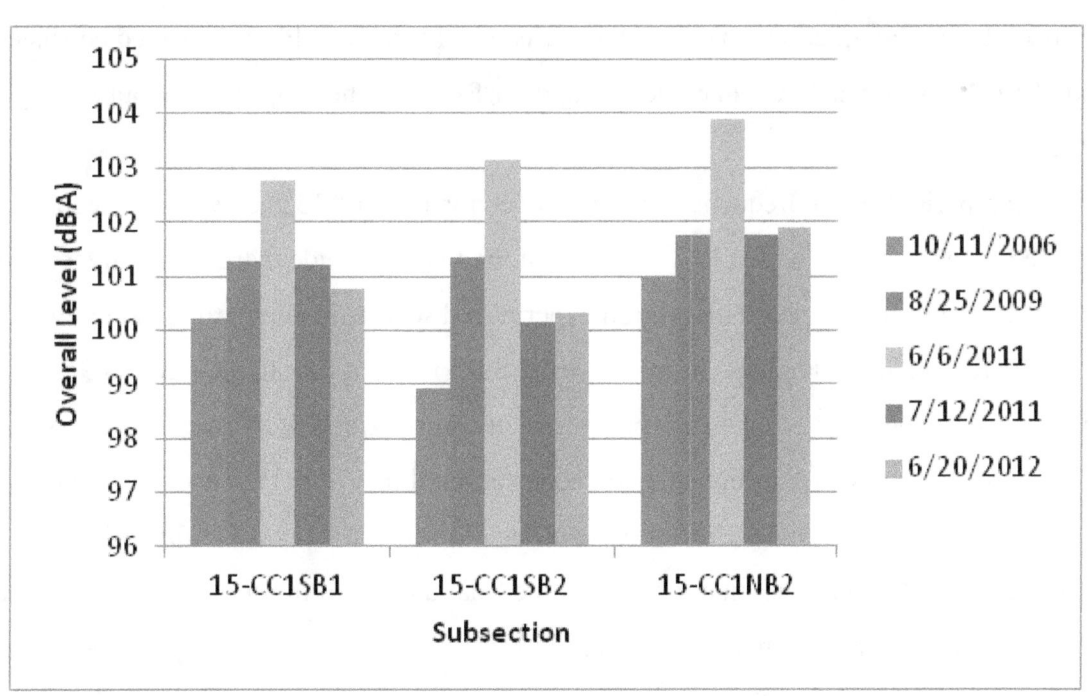

Figure D.26. Overall Noise Level for Section 15, IH-37, Corpus Christi.

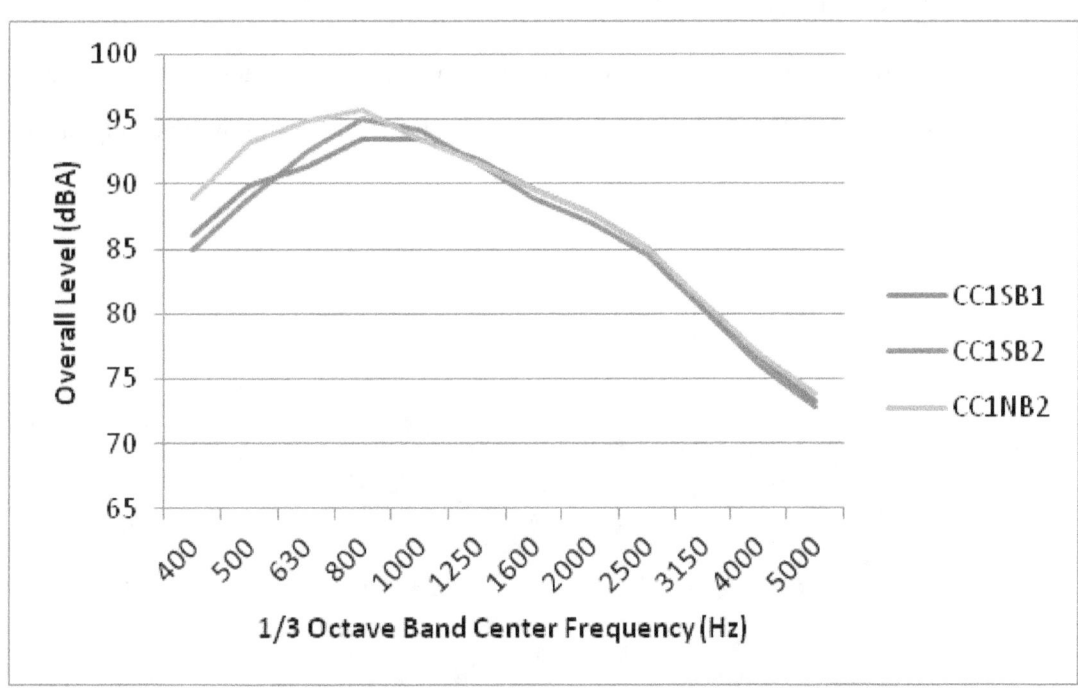

Figure D.27. Frequency Spectra for Section 15 Measured 6/20/2012.

Even though these sections are adjacent to each other, their acoustic performance shows important differences. Section 14, the PG-PFC, was the loudest PFC throughout the study. However, its graph of noise levels over the years, including measurements prior to this project, shows that it was not always loud. It started reaching higher levels with the 2009 measurements, and then kept getting louder. On the other hand, Section 15, the AR-PFC, remained consistent at around 101 dBA, which is lower than the average for PFCs for the project but about average for AR-PFCs.

A comparison of the frequency spectra for sections 14 and 15 shows a significant difference in their shapes. Section 14, the louder of the two by a wide margin, has higher peaks at the lower to mid frequencies. This section experienced severe raveling around 2009, which could be the reason for its loudness in recent years. Such raveling, similar to Section 9 in Waco, could be due to binder aging or the PFC pavement reaching the end of its service life, given that it was constructed in 2004. A recent study on acoustic durability of porous asphalt pavements (Van der Zwan, J.T., *Developing Porous Asphalt for Freeways in the Netherlands*, TR News, January–February 2011) reports that raveling in the final quarter of the service life of PFCs is the main cause of the decline in noise-reducing capabilities.

SECTION 17, IH 35, WACO

The results from these pavements are presented in Figures D.28 and D.29, for the old and the new PFCs, respectively. The frequency spectra for both the old and new PFCs are shown in Figure D.30.

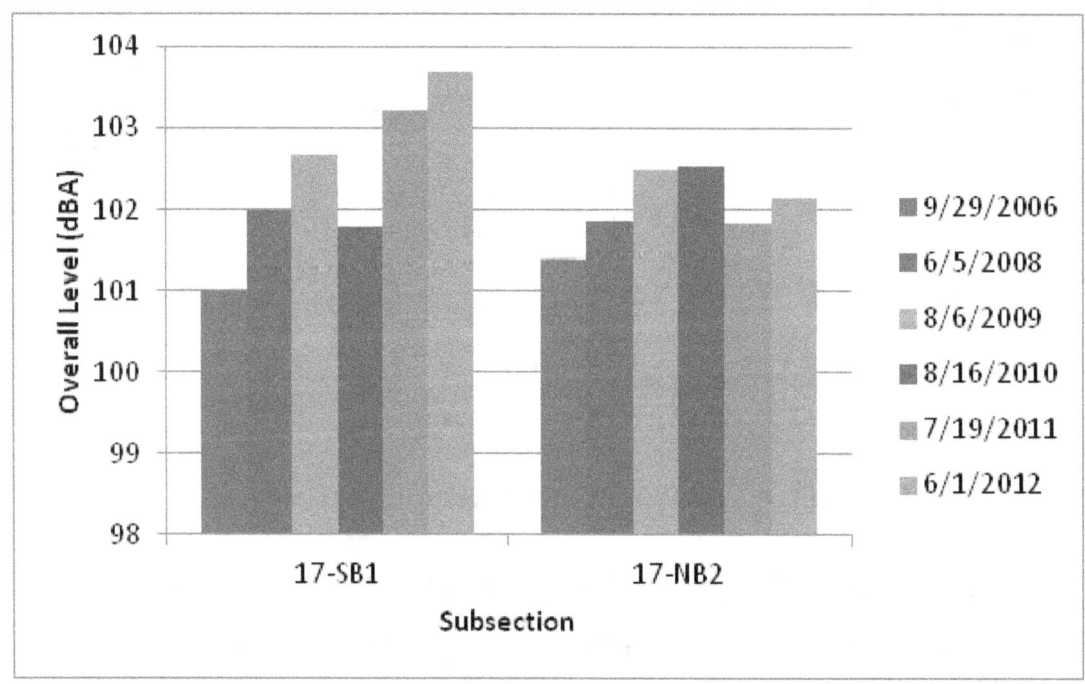

Figure D.28. Overall Noise Level for Section 17 (Old PFC), IH-35, Waco.

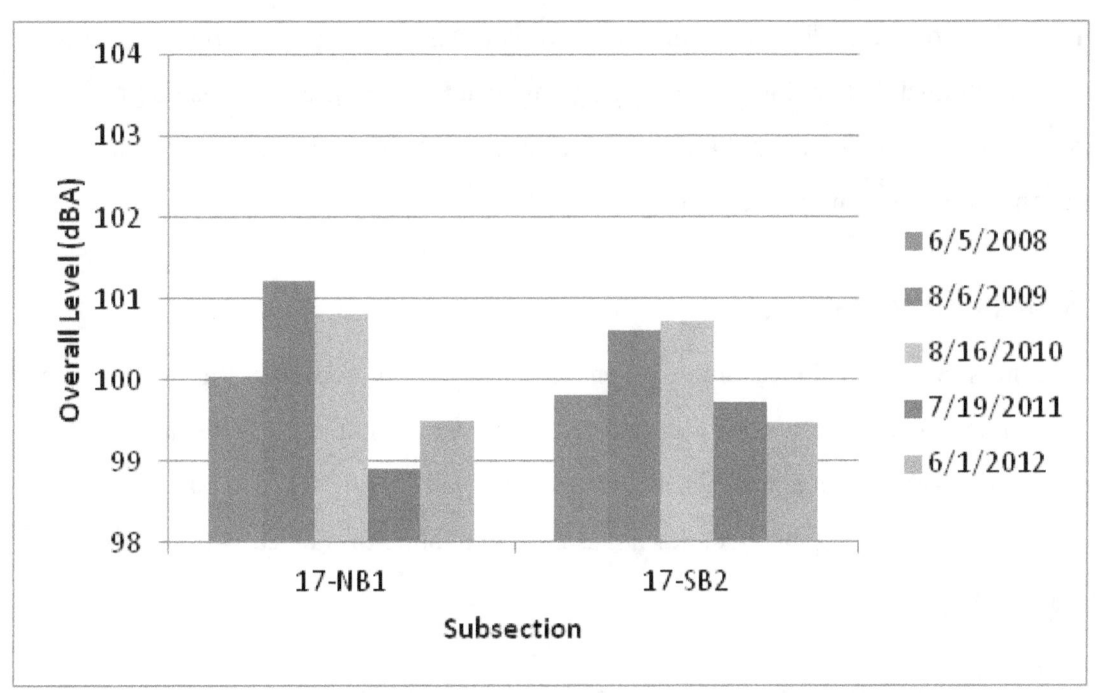

Figure D.29. Overall Noise Level for Section 17 (New PFC), IH-35, Waco.

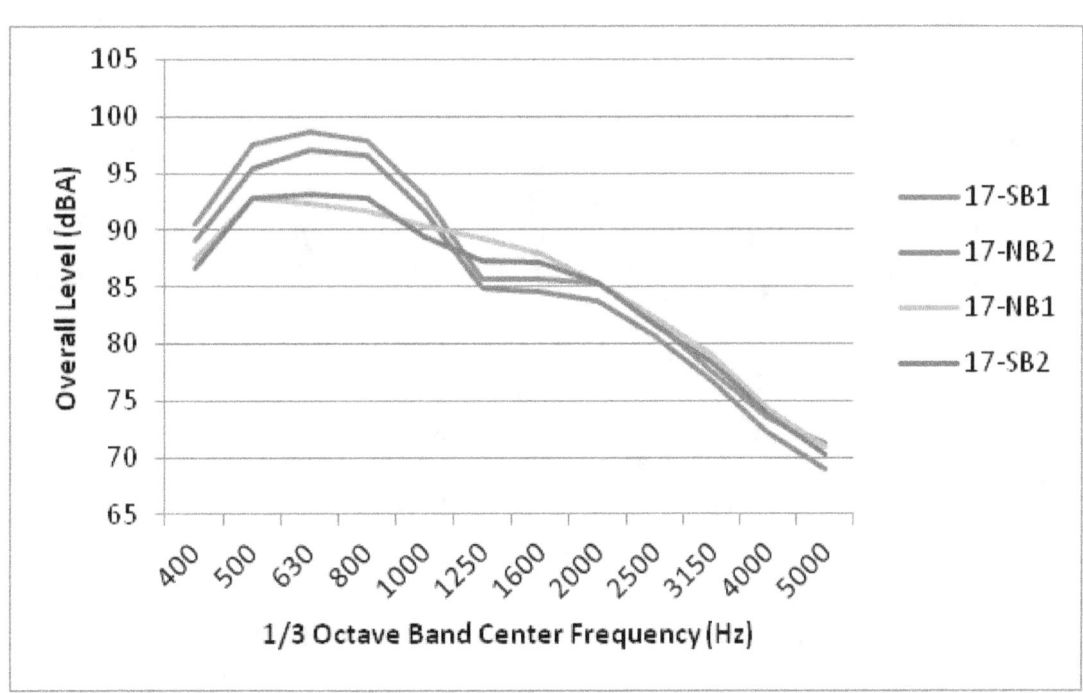

Figure D.30. Frequency Spectra for Section 17 Measured 6/1/2012.

In this case, the new PFC is much quieter than the old PFC. The frequency spectra show that subsections 17-NB1 and 17-SB2, the new PFCs, have flatter curves, especially in the lower frequencies; for frequencies above 2000 Hz, the spectra for the four subsections are very similar.

This section has a very high variability within subsections, which is explained by the fact that there is an old and a new PFC within the same project. Also, other parts of this project have found loss of drainability problems due to clogging, caused by aggregate crushing and subsequent compaction, which judging from the acoustic standpoint, seem to be affecting the old PFC (northernmost end of the project).

SECTION 19, IH 20, TYLER

Results of the OBSI measurements on Section 19 are presented in Figure D.31, and the frequency spectra chart is shown in Figure D.32. The average sound level for this section over the years matches the average for PG-PFCs in this study (102.9 dBA). The pavement looked in very good condition during the visits for noise testing, in spite of the heavy traffic that the highway carries.

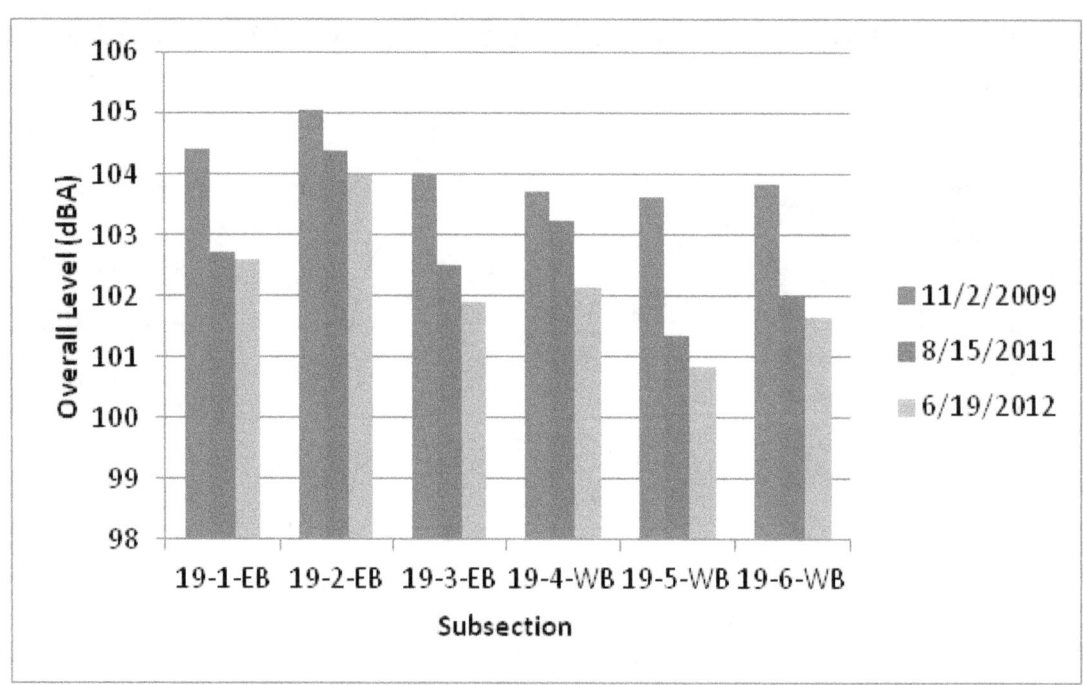

Figure D.31. Overall Noise Level for Section 19, IH 20, Tyler.

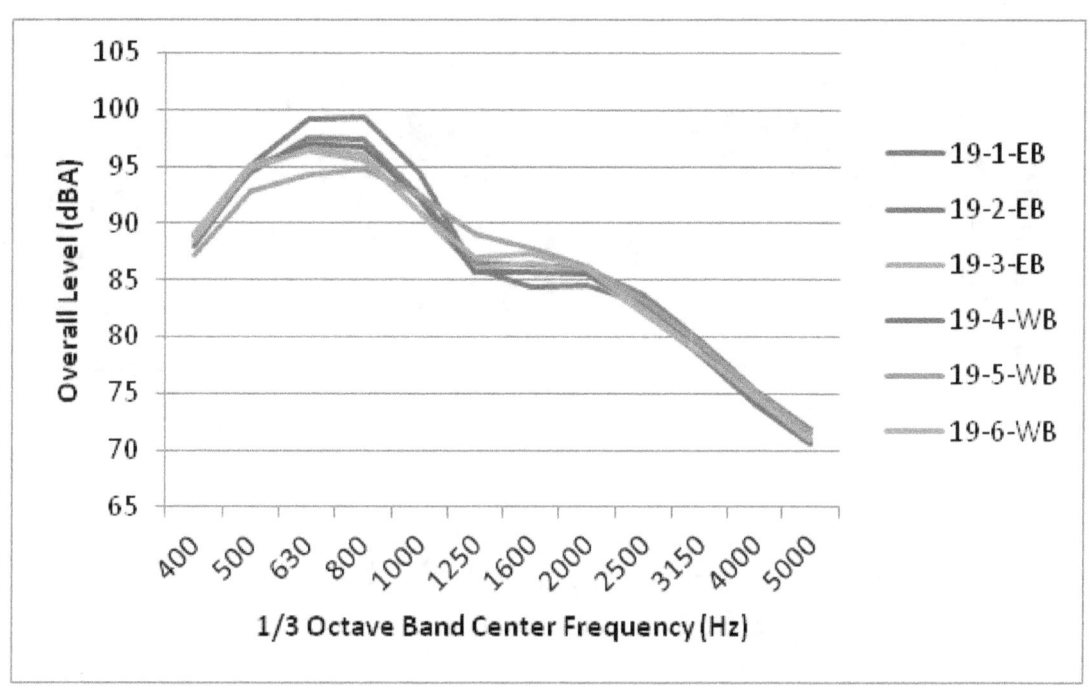

Figure D.32. Frequency Spectra for Section 19 Measured 6/19/2012.

SECTION 20, IH 20, TYLER

Figure D.33 shows the OBSI results from Section 20, and Figure D.34 presents the frequency spectra graph.

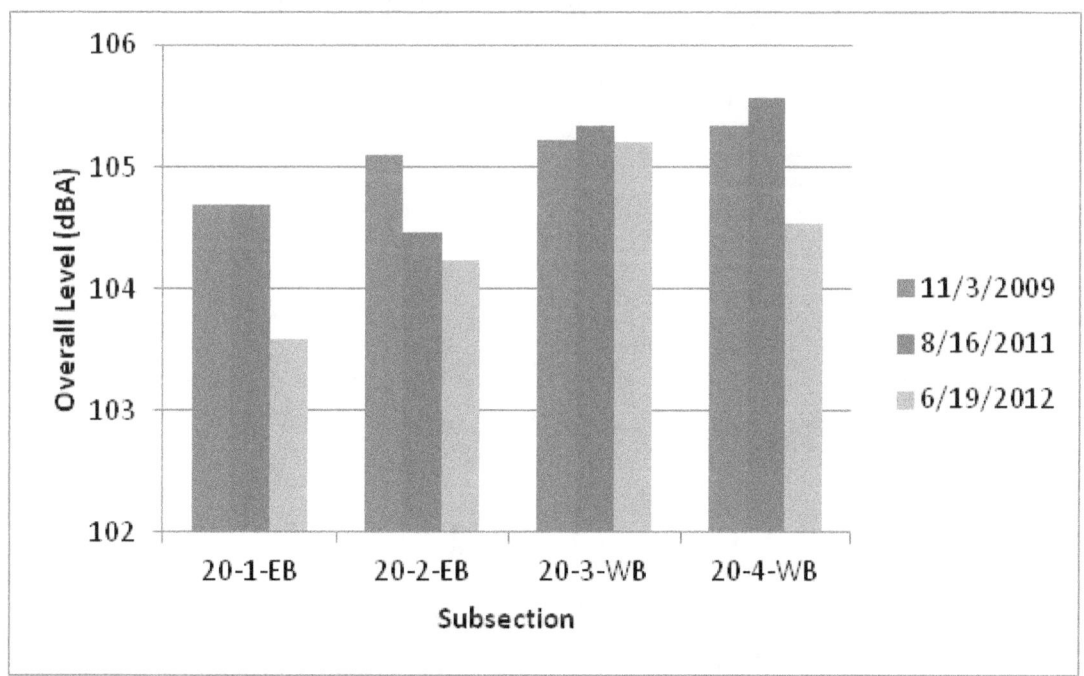

Figure D.33. Overall Noise Level for Section 20, IH 20, Tyler.

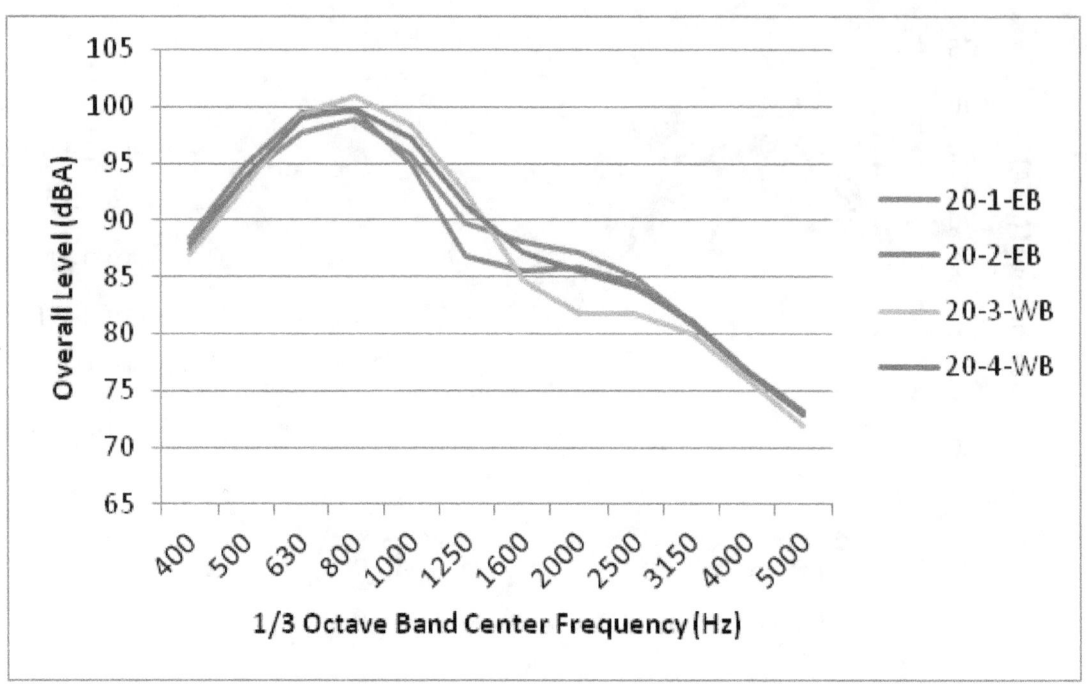

Figure D.34. Frequency Spectra for Section 20 Measured 6/19/2012.

In spite of being the newest pavement in this study, and its apparent good condition, Section 20 in Tyler was consistently loud throughout the years. Its average overall level over time was close to 105 dBA. Section 20 was the second loudest PFC in the study, only surpassed by the aforementioned Section 14 in Corpus Christi caused by raveling.

www.ingramcontent.com/pod-product-compliance
Lightning Source LLC
Chambersburg PA
CBHW080239180526
45167CB00006B/2334